I0100748

GRAPES
A Glossary

The Authors

Dr. S.D. Ramteke, Principal Scientist (Plant Physiology) ICAR-National Research Centre for Grapes, Manjri Farm, Pune-412 307, a pioneer in "Use of bioregulators in viticulture, stress physiology in grapes" and studies on physiological disorders in viticulture, has extended his services on food safety and enhanced export of table grapes for 18 years. Born on January, 18 1966, in a farmer's family at Masalmeta, Bhandara, M.S., Dr. Ramteke passed high school (1982) from Dyaneshwar Vidyalaya Salebhata, and intermediate (1985) from Govind Jr. College, Palandur. He did his B.Sc. Forestry (1989) and M.Sc. Agril. Botany (1991) from PKV, Akola, and worked on the "Physiological basis of variation in yield potential of Chickpea genotypes" at ARS, Hebbali Farm, Dharwad (1991-1995) and received Ph.D (Crop Physiology) from University of Agril. Sciences, Dharwad. Dr Ramteke joined ICAR as an ARS in Plant Physiology during 1996 and worked on crop physiology, bioregulators and water stresses and canopy management. He is life member of five of scientific societies including ISPP, Annals of Plant Physiology, Indian Journal of Horticultural Sciences, Journal of Advance Horticulture, NESA, and published more than 60 research papers, 200 popular articles, covering 5 book chapters/reviews, 6 books. He is recipient of Sanstha Bhushan Award, Rashtriya Shikshan Sanstha Lakhani, Maharashtra in 2001, Scientist of the year 2004 award by National Environmental Science Academy, New Delhi., Abhinav Gaurav 2005' by Abhinav Grape Growers Co-operative Society Agar, Junnar, Pune, Maharashtra, Dr Ramteke visited South Africa and U.S.and worked as Reviewer for vitis vea database, Germany and Referee of the Annals of Plant Physiology, Advances in Plant Sciences, Agricultural Science – bulletin –National Environmental Science Academy, New Delhi and Journal of Maharashtra Agril. Universities.

Mrs. Kavita Y. Mundankar, (Scientist) Computer Applications in Agriculture is presently working at ICAR-National Research Centre for Grapes, Pune as Scientist, Computer Applications in Agriculture since since 2000. She has expertise in development of information systems and has developed information systems related to grapes. She has published various research papers in journals of repute.

Ms. Ujjwala Jape has completed PG in Organic Chemistry, PGDIPR and registered Patent Agent at IPO. She is currently working at ICAR- National Research Centre for Grape, Pune as Research Associate, from 2014. She has experience in chemical research and has good knowledge in Intellectual Property Rights (IPR).

GRAPES
A Glossary

Compiled by
Dr. S. D. Ramteke
Mrs. Kavita Y. Mundankar

Assisted by
Ms. Ujjwala Jape

2017
Daya Publishing House®
A Division of
Astral International Pvt. Ltd.
New Delhi – 110 002

© 2017 AUTHORS

Publisher's Note:

Every possible effort has been made to ensure that the information contained in this book is accurate at the time of going to press, and the publisher and author cannot accept responsibility for any errors or omissions, however caused. No responsibility for loss or damage occasioned to any person acting, or refraining from action, as a result of the material in this publication can be accepted by the editor, the publisher or the author. The Publisher is not associated with any product or vendor mentioned in the book. The contents of this work are intended to further general scientific research, understanding and discussion only. Readers should consult with a specialist where appropriate.

Every effort has been made to trace the owners of copyright material used in this book, if any. The author and the publisher will be grateful for any omission brought to their notice for acknowledgement in the future editions of the book.

All Rights reserved under International Copyright Conventions. No part of this publication may be reproduced, stored in a retrieval system, or transmitted in any form or by any means, electronic, mechanical, photocopying, recording or otherwise without the prior written consent of the publisher and the copyright owner.

Cataloging in Publication Data—DK
  Courtesy: D.K. Agencies (P) Ltd. <docinfo@dkagencies.com>

Grapes : a glossary / compiled by Dr. S.D. Ramteke, Mrs. Kavita Mundankar ; assisted by Miss. Ujjwala Jape.

  pages cm
  Includes bibliographical references.

  ISBN 9789386071491 (International edition)

  1. Grapes—Dictionaries.   I. Ramteke, S. D., 1966- compiler. II.
Mundankar, Kavita, compiler. III. Jape, Ujjwala, compiler.
  QK495.V55G73 2017   DDC 583.8603   23

Published by   :   **Daya Publishing House®**
                                 A Division of
                                 Astral International Pvt. Ltd.
                                 – ISO 9001:2008 Certified Company –
                                 4760-61/23, Ansari Road, Darya Ganj
                                 New Delhi-110 002
                                 Ph. 011-43549197, 23278134
                                 E-mail: info@astralint.com
                                 Website: www.astralint.com

Foreword

India is one among the top ten countries in the world in grape production. It contributes about 2 per cent of world's production, of which 80% comes from Maharashtra followed by Karnataka and Tamil Nadu. The contribution of Indian grapes in the International trade though meager, Indian varieties are in demand in the International Markets.

The compendium of "*Grapes: A Glossary*" is a very good collection and compilation of terminologies often used in grape industry. It will be of great interest to researchers, students, firms and other stake holders engaged in grape cultivation, processing and marketing. The editors have put in hard work to give this compilation a holistic dimension and an integrated approach. I am sure this publication will prove useful to all those interested in the prosperity and promotion of great industry in the country. The glossary is intended to generate the interest and awareness about terminologies used in the cultivation of Table Grapes. It is no doubt, an exhaustive compilation of definitions from different sources and will serve as a ready reckoner for those involved in viticulture.

Dr. M.B. Chetti

Assistant Director General (HRD),
Education Division,
Indian Council of Agricultural Research,
Krishi Anusandhan Bhavan - II,
Pusa, New Delhi – 110 012
E-mail: mbchetti_uas@rediffmail.com

Preface

Grape is commercially important fruit crop which earns lot of foreign currency. It is also a healthy food which contains high vitamins and minerals. It is also a quick source of energy and one of the world's largest fruit crops and most commonly consumed fruits in the world. It contains flavonoids which are antioxidant compounds. Hence, it has got very much importance in human diet.

This book is a compilation of terms related to the science, production and various aspects of grapes. This book attempts to cover words and phrases from all aspects of grapes. The book at present encompasses meaning of more than 1000 terms related to grape arranged in alphabetical order. The meanings are explained in terms of its use in the lifield of Indian Grapes in clear and concise manner. Wherever necessary the terms are explained with suitable photographs to understand the terminology in better prospective.

This book is intended primarily as a reference book for terminology used in grape industry with the aim to provide first-hand understanding of the terms related to various aspects of grapes. The definitions of various terms may need to be known by the researchers, students, several readers and others with interests in table grapes and this compilation will fulfill this need by providing an easy source of reference at one place to terms related to grapes. Also this may serve the purpose of ready reckoner for newly employed personnel. Anyone who is new in the field of grapes or its related area of work, the book will help in quick learning.

Dr. S. D. Ramteke
Mrs. Kavita Mundankar

A

Abiotic: Something which is physical and not biological in nature. Things in nature those are not alive. Abiotic factors such as extremes of heat, light, moisture, lack of nutrients *etc*. has adverse effect on grapevines.

OR

Something which is physical and not biological in nature. Things in nature those are not alive like temperature, humidity, light, rainfall, etc. has an effect on grapevines.

Abiotic Natural Control: Natural non-living factors including (heat cold, rain, drought) which can provide control especially of diseases and pests (insects).

Abscission Layer: A layer of parenchyma cells where separation occurs, that leads to shedding of plant parts such as leaves, flowers or fruits.

http://eviticulture.org/glossary-of-grape-terms/

Absolute Humidity: It refers to the mass of water vapour present per unit volume of space *i.e.*, the density of the water vapour. It's unit of measurement is grams per cubic meter.

http://www.mcgeescrossroadsweather.com/wxterms.php

Absorption: The act of sucking in anything. The movement of a substance across a cell membrane.

Acceptable Daily Intake: The acceptable daily intake (ADI) for grape plant is considered to be a level of intake of a chemicals, nutrients, pesticides, fertilizers; that can be ingested daily over an entire lifetime without any adverse effect.

Accessory Bud: Lateral bud is occurring at the base of a terminal bud or the side of an axillary bud.

Ency. Dic. of Bioinformatics and Biotech. (Set 2 Vols) by D.J. Atary, page no.2; 2004.

Acclimation: The adaptation of a plant or animal to changes in climate or environment.

Acetic Acid: It is a colourless, pungent smelling liquid which is volatile and a primary acid found in wine.

Acidification: It is the method of producing acetic acid by fermentation in wine.

Acetobacter: These are the bacteria's that oxidizes wine alcohol to acetic acid.

Acid Content: In grapes, quality parameters predominantly are TSS and acidity and nearly 90 per cent of the acid content in the grape berries are of tartaric and malic acids.

Acid Soil: A soil which contain pH value less than 7.0. When pH value decreases from 7 to 1, acidity of soil goes on increasing.

Acidity(Grape Berries): It refers to the acid content of grapes which is an important component in the quality and taste of the wine. As the grapes ripen, the level of acidity decreases and sugar concentration level increases. Higher acidity will taste more bitter and sharp and less acidity feel soft and flat.

Acre: An acre is a standard unit of measurement for the area, the size of farms and land estates. 1 Acre =4 046.85642 m².

Acropetal: Acropetal means the route of growth from base to apex of floral buds, tissues or organs ; opposite of basipetal.

Active Ingredient: Substances in a pesticide or any drug that is directly responsible for its desired effect. Usually abbreviated as AI for *e.g.,* active ingredient in the grape juice are flavonols, anthocyanin, flavan-3-ols, procyanidins and phenolics acid.

Adjuvant: Pharmacological or and immunological agent added to a solution that increases or aids in the action of the principal ingredient. In viticulture, sprays, adjuvant, are added to enhance the effect of the active ingredient of the spray.

Adsorption: It is a process in which gas, liquid or solute molecules stick or get attached on the surface of another solid or liquid substance. Grapevines transport certain nutrients by adhesion on the surface of other substances.

Adventitious Bud: A bud that arises at points on the plant other than at the stem apex or a leaf axil.

http://en.wikipedia.org

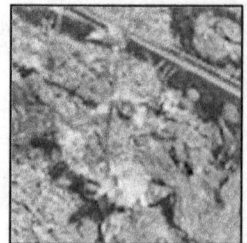

Adventitious Roots: A root that grows from any plant part other than the primary root *e.g.*, from a stem. In grapevines adventitious roots develop from the nodes of a newly planted cutting.

Aerate Soil: It is a process in which air is circulated through soil with small holes to allow air, water and nutrients to penetrate into grape root.

Aerial Root: A root that develops from a point on the plant above the surface of the earth or water, development of root adventitiously above the soil surface *e.g.*, from a stem.

Aeroponics: Aeroponic means growing plants without soil.

AFLP Marker: Amplified fragment length polymorphism is a PCR-based tool used in the research area for DNA fingerprinting.

Agar: Agar or agar-agar is a gelatinous substance, derived from algae. It is a mixture of agarose and agaropectin commonly used to solidify liquid culture media.

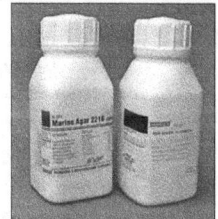

Agribusiness: Business of agriculture related to farming, processing and manufacturing and/or the packaging and distribution of products.

http://www.careerindia.com/courses/unique-courses/what-is-agribusiness-scope-career-opportunities-011796.html

Agricultural Waste: Agricultural waste, which contain both non-natural wastes and natural waste (organic) is a typical process used to describe waste produced on a farm through various farming activities. These activities can include but not limited to pruning, rootstock breeding, pre or post harvesting material, plant culturing.

Agrobacterium tumefaciens: It is rod-shaped, non-sporing, motile, flagellated bacterium responsible for crown gall tumour in plants. Nowadays such bacterium used to transfer genetic material into plants through biotechnology.

Agro biodiversity: Agro-biodiversity is a vital subset of biodiversity. The component of biodiversity that is relevant to food and agriculture production. The term agro-biodiversity encompasses within-species and ecosystem diversity.

Aisle Grape: It is the walkway between or along the rows of plants.

Albinism: Complete lacking of green chlorophyll pigments in plant leaves, resulting from genetic factor.

Alcohol: It is a colourless liquid that is volatile, flammable and that cause intoxication. It is present in wine, beer, spirits. It is used as a solvent in many industries and also used as fuel.

Alkali Injury: Alkali injury to the grape is a problem in the arid tropics of peninsular India, particularly in Maharashtra and north interior Karnataka, where the grape is grown on heavy black soils. This injury is due to high concentration of sodium in the soil.

Alkaline Soil: Soil with pH 8.5 or above and having soluble salts of magnesium and sodium. Development of the grapevine is adversely affected by the high level of sodium in the soil.

Allele: Gene is set of a different allele in that two alleles consider for a single gene. In grapes (*Vitis vinifera* L.) colour associates with allelic variation.

Alley Cropping: Alley cropping is a simple technique that of planting of two rows of grapevines at wide spacing creating alleyways within which other crops of agriculture or horticulture are grown.

Allogamy: Allogamy is one type of sexual reproduction in the plant. Also called cross-fertilization, which is the usual and best-known way of sexual propagation.

Alternaria Rot Or Brown Rot: It is caused by *Alternaria alternata* and mainly occurs near the cap stem (pedicle attachment) as even a slight injury especially bruising of berries caused during improper handling facilitates infection. At first the colour of the rot is tan but later becomes brown. Hence called as brown rot.

Alternative Wine Closures: Other substances used for sealing wine bottles in place of cork.

http://en.wikipedia.org/wiki/Glossary_of_winemaking_terms

Altimeter: A device used by meteorologist to measure the altitude is called as an altimeter.

http://www.ncirossallpointfleetwood.co.uk/weather/terms.htm

Altitude: It is a measure of the height of an object or place with respect to a reference level especially mean sea level.

Amendment, Soil: A soil amendment is any material such as lime, gypsum sawdust, sulphur or soil conditioner, added to a soil to improve its physico-chemical properties, such as water retention, permeability, water infiltration, drainage, aeration and structure.

American Grape: The grape grown in North America.

Amino Acids: It is a class of organic compound having at least one amino ($-NH_2$) and one Carboxylic acid ($-COOH$) group. Amino acids are the building blocks of proteins. They are found in grape.

Ammonium Volatilization: Ammonia volatilization is an important pathway of nitrogen loss from plant soil. Soil conditions including the moisture content, texture, cation exchange capacity, pH and plant residue protect ammonia loss.

Ampelography: This is science in which characteristics of grapevine varieties are identified and described.

http://eviticulture.org/glossary-of-grape-terms/

Anaerobic: It means living without free oxygen. Aging of wine in sealed bottles results in anaerobic changes in it.

Andisols: Andisols are soils formed in volcanic ash and having a highly variable chemical and mineralogical composition like glass and amorphous colloidal materials, including allophane, imogolite and ferrihydrite.

http://en.wikipedia.org/wiki/Andisols

Anion: These are ions that are having a negative charge. (*e.g.*, Cl-) OR An ionic species having a negative charge.

Anther: The part of the stamen that contains pollen.

Anthesis: Process in which flower is fully open and sexually functional.

Anthocyanin: This is a pigment that gives colour to leaves, flowers and fruits and it ranges between red to blue colour.

Anthracnose: This is a disease that occurs primarily during the rainy season . It is present in all grape growing regions of India. It is caused by the fungus *Sphaceloma amprolium (Gloeosporium ampelophagum)*. The perfect stage of the fungus is *Elsinoe ampelina*. Recently, in India the pathogen is shown to be *Colletotrichum gloeosporioides* and not *E. ampelina*.

Anthropogenic: Changes in soils caused by activities done by peoples such as ploughing, fertilizing, and using it for construction.

Anti nutrients: A chemical compound that inhibits normal uptake of nutrients or that interferes with the absorption of nutrients.

Antioxidant: Chemical that prevent oxidation. Chemicals, such as sulfur dioxide which is used to avoid the grape must from oxidizing. Resveratrol is important phytochemical present in grape that also possesses antioxidant property.

Ants Insects: Ants are ecosystem engineers, significantly affecting physical, chemical, and biological properties of the soil. Ants mediated chemical changes of soil are represented mainly by a shift of pH towards neutral and an increase in nutrient content.

Myrmecol News 11:191-199, Vienna, August 2008

Apex: The tip or topmost point of the shoot, leaf, or root, or the blossom.

Apical Dominance: Inhibition of the growth of side branches by the terminal bud of the shoot in a plant. In grapevines, the shoot from the middle to the apex has most growth of leaves and grape bunches.

Apical Meristem: Areas of cell division found at the tip of shoot or root responsible for vertical growth. In grapevines, this area is located at the tip of the shoot.

Appellation of origin: The appellation of origin is a special kind of geographical indication. A geographically based term to identify where the grapes for a table purpose were grown. It consist of a geographical name or a traditional designation used on products. *e.g.*, Nashik table grapes

Application For A Patent: To obtain a patent grant to the invention; it is necessary to file an application with the authorized body (Desired Patent Office) with all important documents and fees. The patent office conducts a required action to decide whether to grant or reject the application on the basis of invention criteria.

Application rate: Mass of pesticide, herbicide, and insecticide active ingredient applied over a specific vineyard area or per unit volume of an environmental component (air, water, soil).

http://www.egeis-toolbox.org/documents/16 per cent 20Abbreviations.pdf

Aqua Ammonia: Solution of ammonia in water is called as aqua ammonia. It is called as a weak base. Its ionization rate is much less in water than strong base such as sodium hydroxide.

Arable Land: Land fit for cultivation.

Arid: Extremely dry condition of climate or lack of moisture.

Arid Climate: The climate in the region that lack sufficient moisture for crop production. In cool regions, annual precipitation is usually less than 250 mm.

Aridic: A soil water regime with extended dry periods.

Aridisols: A soil type which is poor in organic matter and rich in salts; commonly found in deserts.

Aroma And Flavor Compounds: It is a chemical substance that has a smell or odor. Aroma and flavor compounds found in grapes depend upon varieties and area of growing and climatic condition. For, *e.g.,* 'Thompson Seedless' produce fruit with a neutral aroma and flavour. Raisin varieties such as 'Muscat of Alexandria' produce fruit with easily distinguished aromas and flavors.

Aromatized Wine: Wine to which additional flavor is introduced by use of fruits, flowers, herbs and spices.

Ascorbic Acid: This acid is also known as vitamin C. It is found in many citrus fruits including grapes.

Aseptic: Sterile, free from disease causing microorganisms/free of any decay, filth, microorganisms

Asexual Propagation: Plant propagation using the vegetative parts of the plant like shoot, leaf or root.

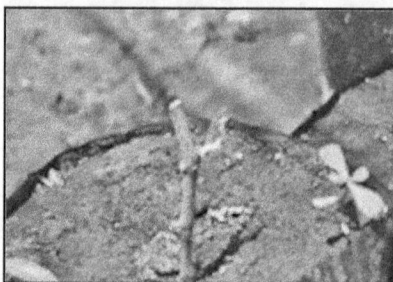

Aspergillus Rot or Black Rot: Black rot caused by *Aspergillus niger*. It is observed in warm to hot conditions. It infects only the injured berries. The area is first turn tan to brown, and is soon covered with a dusty mass of brown or black spores.

Atmosphere: The gaseous portion of the physical environment that encompasses a planet. The earth's atmosphere is divided into troposphere, the stratosphere, the mesosphere, the ionosphere, and the exosphere.

Atmospheric Pressure: The pressure caused by the weight of the overhead atmosphere on a unit area of surface at a given point.

Atrazine (2-Chloro-4 Ethylamino-6 Isopropylamino-S Triazine): It is 2-chlro-4 ethyl amino-6 isopropyl amino – S triazine a widely used selective herbicide for control of broad leaves and grassy weeds in grape field.

Attractant: Chemical or substance intentionally used to attract harmful organisms like the pest for monitoring or other purposes related to control of pest and diseases for *e.g.,* pheromones.

Pure Appl.Chem.,Vol.78, No.11, pp.2075–2154, 2006

Autocidal Technique: The use of insects for self-destruction, chiefly by release of sterile individuals.

http://www.tifton.uga.edu/lewis/glossary.HTM

Autoclave: This is a method of sterilization used in laboratory experiments.

Autoradiography: Autoradiography is the visualization of Radiation consists of X-rays, gamma (g) or beta (b) rays, and the recording medium is a photographic film.12

http://bofduniverse2.blogspot.in/2010_12_01_archive.html

Available Nutrient: It is the amount of nutrient that can be absorbed or taken up by the plants for their growth.

Available Water: Soil available water that absorbed by plant roots. An amount of water available, stored, or released between field capacity and the permanent wilting point.

http://nrcca.cals.cornell.edu/soil/CA2/CA0212.1-3.php

Avenue: Avenue means the broad area at the end of grapevine rows which is usually kept for turning the equipment.

Avicide: Pesticide used for the control of birds.

Avirulent: Unable to cause disease; lacking virulence; nonpathogenic.

http://bugs.bio.usyd.edu.au/learning/resources/PlantPathology/glossary.html

Axil Point: Developing bud point on which a grapevine leaf attached.

B Deficiency: Boron is a vital micronutrient. Vegetative and reproductive development of plant is repressed by boron deficiency. Plasma membrane loses its functional integrity and there are, changes in ion flows and pumping activity.

Backcross: Crossing of a hybrid with one of its parents.

http://en.wikipedia.org/wiki/Backcrossing

Bacterial Canker: Bacterial canker is caused by the bacterium, Xanthomonas campestris pv. viticola. It is a gram-negative rod with round ends, motile by single polar flagellum. Grapevine bacterial canker disease has become a serious disease in the entire grape growing regions of peninsular India *Vitis vinifera*.

http://srpgrapes.com/pest-and-disease.html

Bacterial Diseases: Two bacterial diseases, bacterial blight and black knot have been reported to affect grapes. Disease infect all the aerial part of the vine. The bacterium continues from one season to another through the infected canes, and dormant buds and causes disease under favourable condition.

Bacterial leaf spot: Bacteria *Xanthomonas campestris* cause bacterial leaf spot. The infection is characterized by angular spots due to restrictions imposed by small veins.

Bactericidal: Substance that kills bacteria.

Bacteriophage: A virus that infects a bacterium. Bacteriophage present in wine can attack bacterial starter cultures and inhibit the malolactic fermentation.

Bacteriostatic Agent: A substance that restrains growth of bacteria and bacterial reproduction.

Bacterium: Microscopic single-cell (unicellular) life form. They can cause diseases in plants and animals.

Balance Pruning: It is a pruning method that aims for best fruit quality and vine health by a balance between vegetative and reproductive growth of a grape vine.

Band Treatment: Pesticide or insecticides applied to a linear selected grape strip on or along specific vine varieties rows rather than continuous over the all grapevine area.

Bark, Plant: Woody stem part of vine.

Barrenness of Vines: Barrenness means unable to producing fruit, seed and failure of vines to bear the normal crop and the reduced span of productive life. Causes of barrenness are defective training and pruning practices, inadequate care during the non-bearing period and bud failure.

Basal Bud: A small bud lying at the base of a cane or spur; They do not produce shoots, and if they produce a shoot it is often unfruitful in *Vitis vinifera.*

http://eviticulture.org/articles/page/112/

Basipetal: Leaves or flowers of the plant with progressive development from apex to base (*i.e.*, with the youngest towards the base).

Bat grape damage: Cynopterus sphinex bat is most common and destructive to grapes. They are nocturnal in habit. It plucks the berries from the bunches and suck the juice from berries. Percentage damage goes up to 100 per cent at times. Visits by the bats begin 45 minutes after sunset and foraging continues one hour before sunrise.

Baume hydrometer: It is a measure of the concentration of sugar in juice. It is a relative density (RD) scale used for hydrometer reading used by the French chemist Antoine Baume in the graduation of his hydrometers.

Bees: Bees puncture the skin of ripening berries and feed on the pulp through the hole formed. The damage goes up to 70 per cent at times.

Bench: A natural space of vacant land in between two vine cultivated rows.

Bench Grafting: The process of grafting the scion grapevine varieties on the rootstock, indoors on the bench and not in the field.

Berry: Small sized pulpy, juicy and edible fruit. The grapevine fruit contains grape berries in bunches.

Berry Adherence: It refer tp the attachment between berry and the pedicel. This is a very important parameter that determines the shelf life of the table grapes.

Berry Bloom: This is whitish or ash colour coating of wax on the berries.

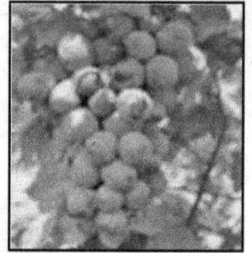

Berry Colour: In grapes, different varieties have the different colour white, red and black. Intermediate shades like pink, brown or greenish black are less desirable. Uniformity and brilliance of the characteristic colour is more important than the intensity of colour.

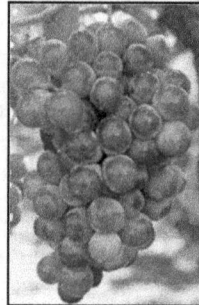

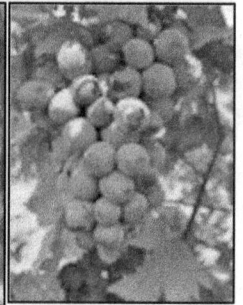

Berry Cracking: This is a physiological disorder of grapes in which berries crack. It occurs due to high soil moisture and high atmospheric humidity.

Berry Drop: This is a detachment of the berries after Harvesting as a result berries falls.

Berry Dry Spot: It is a disorder that occurs in heat sensitive grape varieties due to sunburn in which single berries in a bunch develop sunken, brown and dry spots of irregular shape before softening. Such berries do not ripen. No pathogen is associated with the dried spots of berries.

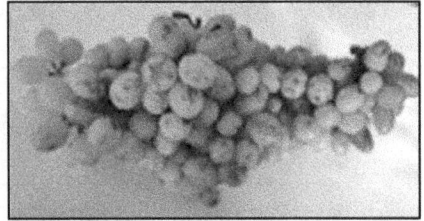

Berry Set: This is one of the important phenological stages of grapevine when the pollinated flowers begin to develop into berries.

Berry Shape: The shape of the berries which is determined by its length and diameter ratio or the ratio of the diameter at the stalk end to that at blossom end. Three distinct shapes - elongated, spherical and oval are observed in grape berries.

Berry Size: It is the lenth or diameter of the berry or both.

Berry Uniformity: It is equality of the berries in a bunch for size, shape and colour.

Bilateral Cordon: They are the two branches from the trunk that are made to grow horizontally in opposite directions of each other in vine training.

Bioaccumulation: Bioaccumulation is process defined as the accumulation of chemicals, mineral, organic compounds in the tissue of grape plant with specific quantity.

Bioassay: Bioassays are procedures that can be used to calculate the concentration of purity or biological activity of a substance such as an enzyme, hormone, phytochemical and plant growth element *in vivo* and *in vitro* condition.

Biochemical Deterioration: Senescence and aging of grape berries caused by the amino acid metabolism particularly of arginine, proline and glutamic acid.

Bioconcentration: It is an ability of a microorganism to store material in the environment.

Biocontrol: Reduction of pest inhabitants by natural enemies.

Biodegradable: Any substance that can be decayed by biochemical processes of bacteria or any other living organism.

Biodynamic Farming: Biodynamic farming integrates agricultural, bio-ecological scientific knowledge into vinery rotation, manure production, plant diversity, homeopathic sprays and soil and animal practices.

http://www.ceago.com/biodynamic/

Biofertilizers: Biofertilizers are organisms that enrich the nutrient quality of the soil. The primary sources of biofertilizers are bacteria, fungi, and cyanobacteria (blue-green algae). For, *e.g.*, Vitec - Grape Biofertilizer; specially developed for the grape crop. This fertilizer contains different strains of a useful microorganism.

Biological Half-Life: The biological half-life or elimination half-life of a substance is the time it takes for a substance into plant or biological system to be reduced to one-half of its initial value to lose half of its physiologic or radiologic activity.

Biomass: Biomass is carbon-based mixture of organic molecules containing nitrogen, oxygen, alkali and alkaline earth metal. Plant material, vegetation and agricultural waste used as a fuel or energy source.

Biome: 1. A biome is a large geographical area of distinctive plant and animal groups, which are adapt to that particular environment.

http://www.blueplanetbiomes.org/world_biomes.htm

2. A biome is climatically and geographically defined by distinctive communities of plants, animals and soil organisms supported by similar climatic conditions.

http://minecraft.wikia.com/wiki/Biomes

Biopesticide: It is a pesticide made by using naturally occurring substances or organisms. *e.g.* Neem tree extract and some seaweed extract are used as pesticides.

Bioremediation: It is reclamation or improvement in the soil fertility status by various means of decontamination or restoration of polluted or degraded soil and water by using the biochemical degradation or other microbial activity.

Biosenser: A biosensor is an analytical device that converts a biological response into an electrical signal.

http://www1.lsbu.ac.uk/water/enztech/biosensors.html

Biosensing: Technology for the detection of a broad range of chemical and biological agents, including bacteria, viruses and toxins, in the environment and humans.

Principles of Biochemistry and Biophysics; By Dr. B.S. Chauhan page 780; 2008

Biosphere: The zone between the earth and the atmosphere within which most terrestrial life forms exists.

Biotechnology: A set of biological techniques developed through basic research and now applied to research and product development. Biotechnology refers to the use of recombinant DNA, cell fusion, and new bioprocessing techniques.

http://ghr.nlm.nih.gov/glossary=biotechnology

Biotic: Concerned with living organisms.

Bird grape damage: Birds feed on the grape berries directly causing a very heavy loss. Fruit loss ranges from 20 to 50 per cent. Birds prefer more of coloured varieties than white varieties. More damage by birds is noticed in grapevine trained with Head system than in Bower system.

Bird Net: A form of bird pest control in which a net used to prevent birds from feeding on grape berries. Properly installed net in grape vineyard gives protection and saves crop from birds.

Black Soil: Black soil is very argillaceous, very fine - grained and black and having a high percentage of calcium and magnesium carbonate. They are very tenacious of moisture and exceedingly sticky when wet.

Bleeding: It is a process in which sap comes out from the wood of the plant after injury or cut due to the pressure of sap within the tissues that conduct water and sugars around the plant.

Blind Buds: Buds that fail to produce leaf, shoot or flower; nodes on the canes of grapevine that do not develop buds.

Bloom: 1. The time when the calyptra's (flower caps) separates from the flower, also known as anthesis.

2. The waxy coating on the surface of the grape berries.

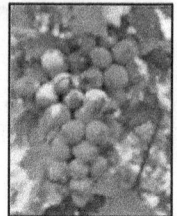

Blossom-End Rot: It is a physiological disorder in tomato. It appeared sporadically in Anab -e-Shahi grape in Hyderabad region during the late sixties. In grapes, it is characterized by the appearance of a small spherical brown rot surrounded by a water-soaked area at the blossom-end of the berry. On no other parts of the berry such spots develop. Brown spot gradually enlarge with the water-soaked tissues around. The affected skin and pulp become soft, dry and sunken. No pathogen is associated with this disorder.

Bordeaux Mixture: It is the name of a fungicide that is used to control fungal diseases in grapevine.

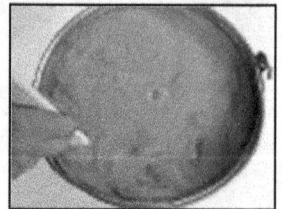

Boron: Boron is essential for cell formation in plant during fruit set. A deficiency of boron causes a low growth of fruit set.

Botrytis Rot: Botrytis is a fungal disease that decomposes grape bunches and sometimes shoots and leaves. Botrytis rot, also known as grey rot, is characterized by the decay and rot of berries with a greyish olivaceous growth of the fungus. The affected berries turn soft and brown and generate an unpleasant smell.

Botrytized Grapes: Grapes that have been rotted by fungus *Botrytis cinerea*.

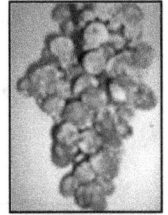

Bower: It is a type of training system for grapevines, suitable for vigorous varieties with a high degree of apical dominance.

Branch Wilt: It is fungal diseases caused by *Hendersonula toruloidea*. Cracking, shredding and peeling of bark are the first symptoms. Later, black cankerous deposition developed downward into the main branch, and soon the entire vine dried up.

Brassinosteroid: It is a new class of chemical that increases the size of the berry. It is polyhydroxy steroids type compound. Essential for normal plant growth, development and increases the size of a berry. This compound also help to promote ripening.

Brix: A measure of sugar content in grapes expressed in degrees. Each degree of Brix equals 1 gram of sugar per 100 grams of the grape juice. Brix is measured at harvest.

http://www.seslisozluk.net/danca/Brix

Brix Yield: This is a quality yield parameter that reflects the yield of quality fruit in brix yield. It is calculated by the formula brix yield(tonnes/ha)=Fruit yield(tonnes/ha) *TSS(°B)/100

Broadleaf: Broad and flat shape leaf rather than slender or needle-like leaf.

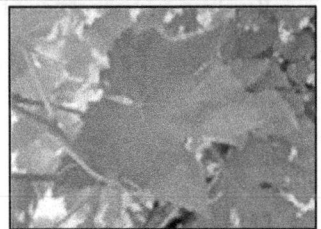

Broad-Spectrum Pesticide: Chemical or substance that kills a wide range of pest species.

Broth: A liquid medium containing the different types of macro-micro elemental nutrient, which is utilized by bacteria and other microorganisms for growth in the cultural environment.

Brown Leaf Spot: *Cercospora viticola* causes the fungus disease. Leaf spots caused by this disease are circular, rarely irregular, dark brown with an grey ashy center.

Bud: Undeveloped primordial shoot that appears as small swelling on the node of a grapevine shoot or cane from which a new shoot develops.

Bud Scales: Scale-like leaves that acts as protective coverings over buds.

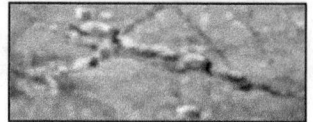

Bud Sprout: A bud is an undeveloped shoot, leaf or flower, in the shape of a small knob.

http://www.italki.com/question/125746

Bud Stick: It is the section of a shoot or cane that is used as a source of buds for budding.

http://eviticulture.org/glossary-of-grape-terms/

Bud break: Bud break refers to onset of growth from a bud. New growth by vines starts with bud break.

Budburst: Emergence of new leaves from grape buds.

Budding: Refers to grafting of vine with buds usually onto a
   rootstock.

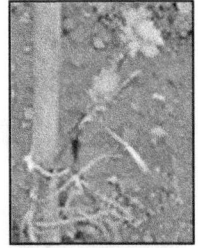

Bung: Bung is a cork or stopper made of rubber or glass which is used to close up a
   container, such as a bottle, pipe or drum.

Bush Training: It is most used vine training system where the vines are kept as
   individual, and vines plant not supported by a trellising system. This is an easy
   method to maintain, requires minimal pruning.

Buttresses: These are the kind of trellis which supports to the trees/trunk in the
   rainforest areas vine.

4-Chlorophenoxyacetic Acid/4 CPA: It is a 4-Chlorophenoxyacetic Acid, plant hormone called auxin; used in the food industry as well as a plant growth regulator.

Calcareous Soil: Saline soil has high pH (alkaline) containing Ca and Mg carbonate with high soil organic matter.

Calcium: Calcium is an essential plant nutrient. It is essential to promote the growth of cell wall and membranes. Sometimes an excess of calcium retards some plant diseases.

Callus: A tissue that grows on or around an injury or damage surface of a plant to heal it.

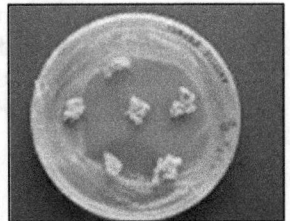

Calvin Cycle: In plant, Calvin cycle is used directly to fix carbon dioxide. Those plants that utilize just the Calvin cycle for carbon fixation are known as C_3 plants.

Calyptra: The membranous hood or cap is covering the calyx.

Calyx: All the sepals are collectively called as calyx. It forms the outer floral layer.

Cambium: Meristematic tissues that contain Cambial cells; which divided to produce secondary xylem cells at centre and secondary phloem cells toward the outside of vine stem.

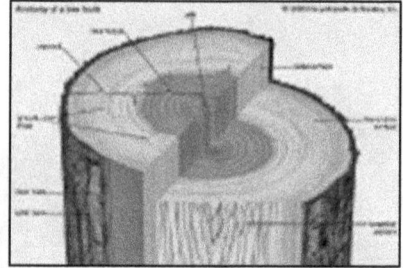

Cane: A mature shoot called cane. Canes become "older wood" after some year. The pith of a cane is also much easier to distinguish than in older wood.

Cane Density: This indicates the number of canes per unit area.

Cane Immaturity: The cane does not turn brown in colour after 150 days after back pruning and they remain green at the time of fruit pruning is called as cane immaturity.

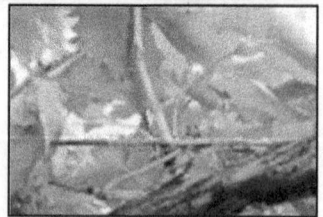

Cane Training: The process of making shoot growth of vine to a desired structure.

Canker: It is an infected wound in the cambium of plants that does not heal over, or it heals over very slowly, usually caused by a fungal pathogen.

Canopy: The foliage of a crop; in grape vineyards the canopy is said to be closed when plant growth of adjacent rows shades row middles and canopy is said to be open when direct sunlight penetration between rows is present.

http://eviticulture.org/glossary-of-grape-terms/

Canopy Management: The action of manipulating/managing shoots, leaves and fruit for the improvement of vine and fruit quality.

Cap Fall Stage: This is the natural phenomenon in which caps falls after Berry setting.

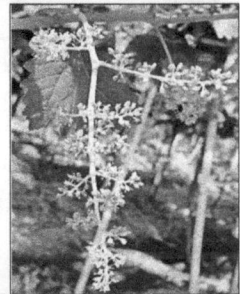

Cap Stem: Also known as pedicel. It is a small stem through which individual grape berries are attached to the cluster.

http://eviticulture.org/glossary-of-grape-terms/

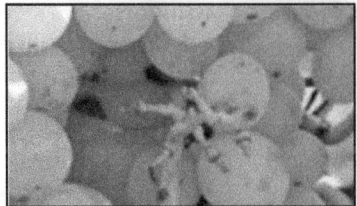

Capacity: For grapevines, it is the total growth and crop that a vine is capable of producing in a year.

http://eviticulture.org/glossary-of-grape-terms/

Capital Investment: Money spent on equipment, stock or on the improvement that has a life of more than one year and which adds to the productive capacity of an enterprise.

Carbendazim: Carbendazim is a broad-spectrum systemic fungicide. It used for the control of a wide range of fungal diseases such as spot, powdery mildew, scorch, rot, blight, *etc.*

http://pdf.usaid.gov/pdf_docs/PA00K8NJ.pdf

Carbon Dioxide (CO_2): Carbon dioxide is constantly produced and absorbed many microorganisms, plants, and animals. More CO_2 is removed from the atmosphere and stored in plants and trees. In brewing, alcoholic fermentation is the conversion of sugar into carbon dioxide gas (CO_2).

Carbonate: Simplest oxo-carbon anion; At the time of bottling wine CO_2 is captured and carbonates the wine, due to this bubble produce that make sparkling wine so delightful.

Carbonic Acid: An inorganic compound with the formula H_2CO_3. It used to carbonate sparkling wines.

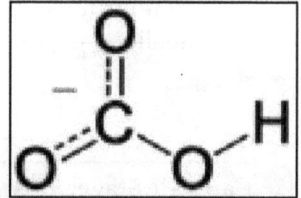

Carbonic Maceration: The process of separating unwanted particles (such as dead yeast cells or fining agents) from the wine by use of centrifugal force.

http://en.wikipedia.org

Casts, Earthworm: Earthworm is a cylindrical tube warm. Improving soil fertility by converting organic matter into rich humus typically active only if water is present.

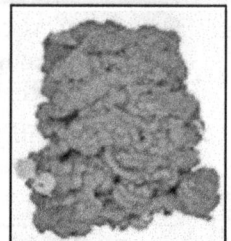

Cat Clay: It is also called acid sulfate soil; contain iron sulfide and their oxide. Acid sulfate soils are widely occurred near coastal areas.

Catch Wire: A wire that is used for shoot growth placement in the vertical shoot positioning training system.

http://eviticulture.org/glossary-of-grape-terms/

Caterpillar: Caterpillar is insect pest have soft bodies that can grow rapidly between molts. Several caterpillars are known to damage the bark, leaves, flower panicles and bunches leading to a severe loss to grape growers.

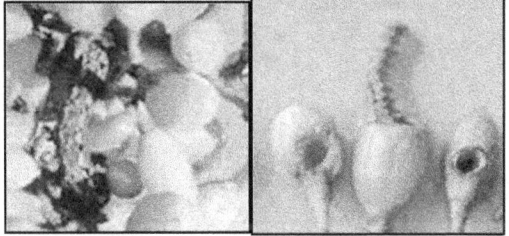

Cat-ion: The term is used in chemistry for positively charged ions. (*e.g.*, Na^+, Ca^{++}, K^+).

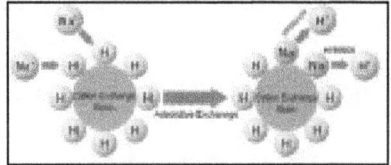

Causal Agent: The organism which causes disease.

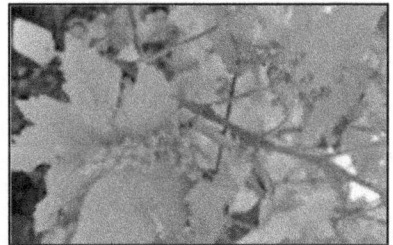

Cavitation: Rapid formation of bubbles in the grapevine xylem is called cavitation's.

Cellulose: Cellulose is an organic polysaccharide with a long chain that form outermost cell wall of the plant. Cellulose is the most abundant organic polymer on Earth.

Celsius Temperature Scale: This is a unit for measurement of temperature scale where water at sea level has a freezing point of 0°C (Celsius) and a boiling point of +100°C. More commonly used in areas that observe the metric system of measurement. Created by Anders Celsius in 1742. In 1948, the Ninth General Conference on Weights and Measures replaced 'degree centigrade' with 'degree Celsius.'

http://xoap.weather.com/glossary/c.html

Certified Planting Stock: Planting material that is certified by authorized bodies as true-to-type and disease and insect free.

Chafer beetle: They are hard bodied insects with horny front wings having different colours. Both adults and larvae are destructive. Adult beetles visit vineyards between 7-11 PM and they start feeding from the periphery of leaves and tender shoots at night.

Chelate: Complex compounds are consisting metal ion and a chelating agent. Chelates play important roles in oxygen transport and photosynthesis.

e.g., Chlorophyll, which is responsible for the green colour of plant leaves, absorbs the light energy.

Chemical Injection: It is a method of application of fumigants or injection of a chemical into the affected plant during different stages helps to kill the larvae.

Chemical Thinning: GA$_3$ is being used for berry thinning in grapes. Single application of GA$_3$ @40 ppm at 50 per cent flowering stage increases the berry thinning.

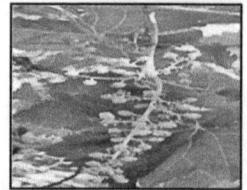

Chemigation: Chemigation is the injection process of a chemicals such as nitrogen, phosphorus or a pesticide into irrigation water and applied to the land using the irrigation system. Chemigation always to benefit your crop production.

Chemodenitrification: It is non-biological processes leading to the production of gaseous forms oxides of nitrogen.

Chicken and Hen: It is a physiological disorder where some berries are bigger in size and some are very small in size called as chicken and hen. It is disorder in a cluster where many short berries surround a bold berry. The bold berry is compared to the hen, the shot berries to chicken. This is mainly due to impaired fertilization and the growth of many berries without embryo formation by exogenous application of growth promoters and is normally due to deficiencies of zinc and boron.

Chilling Requirement: This is a requirement of cold temperature to complete the life cycle of fruit bearing tree or grapevine to flower.

Chip-Budding: It is one of the easier forms of grafting. In which vine propagation by cutting the xylem and phloem of a grape bud into a tiny wedge shape and then inserting into the rootstock of an existing root system.

Chitin: This is a proteinaceous substance tough, protective, semitransparent substance, primarily a nitrogen-containing polysaccharide, forming the principal component of arthropod exoskeletons and the cell walls of certain fungi. A colourless polysaccharide that serves as the major fibrous component of the insect cuticle or integument.

Chitinase: These are the enzymes present in grape juice. The activity of these enzymes increases at the onset of ripening in grapes.

Chlorophyll: Green pigments found in all green plants. With the help of chlorophyll, pigment plant can make their food.

Chloroplast: This is a cell organelle consist of chlorophyll disk-shaped structure in plant and algal cells that contains chlorophyll and is the site of photosynthesis. It is specialized subunits within a cell.

Chlorosis: It is one disease condition in which yellowing of leaf tissue due to a lack of some nutrient like manganese, zinc or iron causes chlorosis.

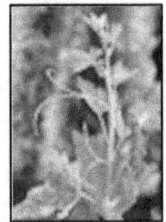

Chronosequence: Soils chronosequences are valuable tools for investigating rates and directions of soil and landscape evolution.

Catena (Impact Factor: 2.48). 06/1998; 32:155-172

Cladosporium Mould: Cladosporium genus is one of the most common environmental fungi. Cladosporium are found in many different types of soil as well as on all kinds of senescing and dead plant matter.

Cladosporium rot or green ball rot: Cladosporium herbarum causes this rot. Infection takes place through the skin, either in the vineyard or storage and may occur at a temperature from 1-30°C. The decay penetrates through the pulp sometimes up to the seeds.

Clay Films Soil: Coatings of oriented clay on the surfaces of peds and mineral grains and lining pores, also called clay skins, clay flows, illuviation cutans, or argillans.

http://eusoils.jrc.ec.europa.eu/ESDB_Archive/glossary/Soil_Terms.html

Climate: The climate is the average weather in a place over many years. Climatic condition changes but required many years.

Climatology: The study of climate-prevailing atmospheric condition of humidity, temperature, wind, *etc.*

Clone: Plant that is derived by asexual reproduction from a single parent plant and is therefore of the same genetic constitution.

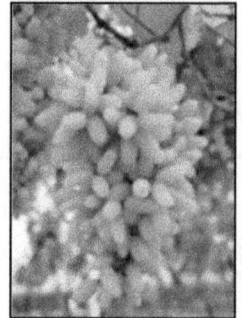

Glossary: Irrigation, Drainage, Hydrology And Watershed Management by R. T. Thokal, A.G. Powar, D.M. Mahale, R.T. Thokal, D. M. Mahale, A. G. Powar; page No.49; 2004.

Cluster: It is a bunch on the grapevine.

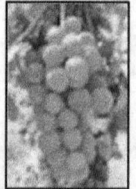

Cluster Cane Ratio: This is a ratio of no. of clusters per cane.

Cluster Stem: The stalk that attaches the grape bunch also called as cluster to the shoot.

http://eviticulture.org/glossary-of-grape-terms/

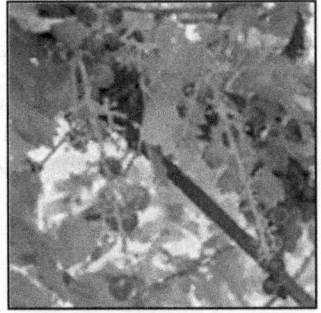

Cluster Thinning: It is also known as bunch thinning. It is the process by which some grape clusters are removed to reduce the crop load on the grapevine so as to increase the quality of the produce.

Coarse Texture: The texture exhibited by sands, loamy sands, and sandy loams except very fine loams.

Cold Wave: Rapidly drop in a temperature level.

Colonization: Colonization of a host results spreading of similar species of the pathogen on or in the infected plant to initiate colony.

Colour Development: The basic colure of grape berry, namely blue, pink, purple and black are due to anthocyanin pigment. Colouring of grapes depends upon the level of anthocyanin in the berries. A several environmental factor like temperature, light and soil moisture influenced on pigment development.

Commercial Fertilizer: Fertilizer is available on a commercial basis in the market.

Compatibility Graft: The ability of a scion variety to grow on a rootstock variety when grafted.

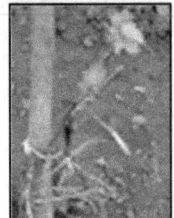

Complex Fertilizer: The commercial fertilizers containing all three primary nutrients (N-P-K) are called as complex fertilizer.

Compost: Compost is usually made by collecting organic matter as decomposing leaves, manure, kitchen scraps, grass clippings, and vegetable peels. It is a rich source of nitrogen, phosphorus, potassium, carbon, and calcium and used as fertilizer and additive.

Condensation: This is a formation of liquid from the gas through the process by which water vapor undergoes a change in state from a gas to a liquid. It is the opposite physical process of evaporation.

http://www.nkhome.com/kestrel/weather-resources/weather-glossary/

Conservation Of Soil: Soil conservation means, keeping its power active to produce food. Without adding soil food, it will have no power to create and will die.

Integrated Organic Farming Handbook; By Dr. H. Panda; page.16.

Contact Herbicide: Non-systemic herbicides that kill plant parts by its contact instead of its absorption by plant parts and its translocation.

Contact Insecticide: Non-systemic insecticide that kills insect pests by its contact.

Contact Poision: Pesticide (herbicide) that causes injury to the target plant tissue to which it applied through physical contact rather than through ingestion or inhalation.

Contaminant: A substance that cannot be readily removed from vineyards.

Copyright: The legal right granted by Indian Copyright Office to an author, composer, playwright, publisher, or distributor to exclusive publication, production, sale, or distribution of a literary, musical, dramatic, or artistic work.

Cordon: Grapevine arms originating from the trunk and usually trained along the wire.

Cork: A bottle stopper made from the thick outer bark of an oak tree.

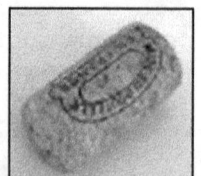

Corolla: It refers to all the petals in a flower as a unit.

Coulure: Grapevine disorder in which flower bud or flower drops off. It is also called as Shelling. Sometimes the flowers that fail to develop into berries fall down from the cluster also called as shatter that leads to loose and poorly shaped bunches.

Cover Crop: The crop planted between main crops basically to protect the soil, control weeds and improve its properties.

CPPU/N-(2-Chloro-4-Pyridyl)-N$_2$-Phenylurea: It's common name is chlorfenuron and the chemical name are N-(2-Chloro-4-Pyridyl)-N$_2$-Phenylurea. It is bio-regulator used alone or in combination with other growth regulators to increase the size of berries.

Cradle: It is agriculture device used for to reap a crop at the time of harvesting.

Crawler: Initial development stage of an insect in which it has legs and the ability to move; usually refers to the first growth stage of an insect.

Cream Tartarates: It is a byproduct of winemaking process also known as potassium hydrogen tartrate. Tartrate prepared from grape stems. Tartarates may be seoarated as the cream of tartar of calcium tartarates. Tartarates separate from new wines because they are less soluble in alcohol than in non-alcoholic grape juice.

Crisp (Noun Crispness): Firm and fresh; not soft or wilted. Specially grapes berry of some varieties are very crispy.

Crop Cost: These crop costs refer to the value of plant, manure, fertilizer, pesticides and irrigation.

Crop Load: This is an indication of fruit production load of fruit production on an individual plant. In grapevine it is determined by a number of buds per vine.

Crop Residue: The portion of a plant or crop that is left in the field after harvesting fruit vineyards.

Cross, Plant: It is a plantlet/species produced through hybridization

Cross-Arm Trellis: It is a kind of trellis used for wine training processes. Trellis design for dynamic canopy manipulation.

Crown Suckering (Thinning): In this process dead, rubbing, split, defective branches removing for increasing airflow and health of the canopy. It also helps to improve fruit size and fruit quality of vine.

Crown, Plant: All above-ground parts of a plant.

Crown-Cap/Crown Cork: A bottle cap.

Cryogenic Soil: Soil that has formed under the influence of cold environment.

Cultivator: Cultivators are either spring tined or rigid tined. It used for heavy work for breaking large clods.

Culture: A particular kind of organism, cells, tissue, organism, or virus growing in a laboratory medium with the controlled condition.

Cumulative Effect: Overall change which occurs after repeated action doses of a chemical substance or radiation on applied plant area.

Cynobacteria: Cyanobacteria can be found in almost every terrestrial and aquatic habitat. Cyanobacteria include unicellular and colonial species. These bacteria play important role in nitrogen fixation.

Cytology: Cytology is a branch of life science, which deals with the study of cells in terms of structure, function and chemistry.

D

Dagger nematode: Dagger nematodes are ectoparasites and feed on succulent tissues of young roots of vinery. Dagger nematode feeding on roots result in terminal swelling, cessation of root elongation, distortion due to malformation of rootless and also stunting of vine shoot and root.

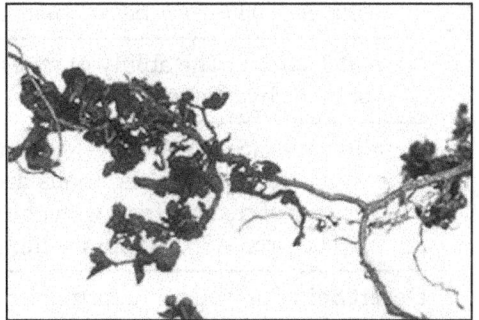

Dalapon (2,2-Dichloropropionic Acid): 2,2-Dichloropropionic acid is a selective herbicide for the control of annual and perennial grasses. It is used to control the type of weeds in between vine rows in a vertically trained vineyard of India.

Dead Arm: Dead arm is the symptomatic expression of the disease/disorder caused by various fungi like *Botrydiplodia theobromae, Macrophomina phaseolina* and *Pestalotiopsis viticola* found in India. Another caused of dry arms moisture stress and dry weather.

Decay/Rot/Putrefy/Spoil: During storage destruction or decomposition of grapes because of bacterial or fungal action or growth.

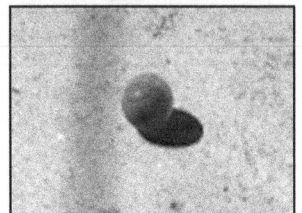

Deciduous: Seasonally plant that dropping off or shedding or losing or falling off all of their foliage at least one time in a whole year.

Deflation: The removal of fine soil particles from the soil by the wind.

Defoliation: The loss of or fall off green leaves from the plant.

Degree Day: It is a heating or cooling measurement phenomena, agricultural activities like pest control, planting or crop harvesting all are a plan by using degree days calendar.

Dendrometer: Plant growth, height and diameter measuring instrument

Dentate: Leaf is having a toothed margin or tooth-like projections or processes.

http://www.thefreedictionary.com/dentate

Dermal Toxicity: The ability of a pesticide or other chemical to poison people or animals by contact with the skin.

Desalinization(Soil/Water): 1- Removal of some extra per cent of Salt from the soil by using leaching process. Soils are considered saline when the EC is >4. The process of creating fresh water by removing saline (salt) from bodies of saltwater 10,000 ppm – 35,000 ppm is high salinity.

Desert Soil: This soil contains mantle of blown sand which combined with the arid climate. The most predominant part of soil is quartz with calcareous grain.

Desiccant: In agriculture, a substance used for drying up plant stems and foliage to facilitate their mechanical harvesting.

http://sis.nlm.nih.gov/enviro/iupacglossary/glossaryd.html

Desiccation: Desiccation means "excessive loss of moisture from berries." It works better with some condition but sometimes it is un useful. Spoiling of grapes due to the high loss of moisture/water that affect on the storage condition of berries.

Destemming: The process of separating stems from the grape berries from the stalks.

Deutonymph: The third stage of a mite or third instars of a mite.

Dew: A Phenomenon in which atmospheric moisture condenses at a high rate than at which it can evaporate; resulting into small water drops/droplets on the surface of an object.

Diagnostic: The identification, determination and detection of the nature of an illness or other problem/condition or disease by examination of the symptoms in the test target.

Diaheliotropism: The movement of leaves or other plant organs in such way that their dorsal surface remains perpendicular to the rays of light; in such position plant photosynthesis rate increases.

Dieback: Spontaneously death of the young shoots, tips, roots, branches or whole plant caused by injury or attack by fungi or bacteria or some environmental conditions.

Diffusion: The movement of molecules or ions of a solute or solvent from plants higher concentration to lower concentration is called as diffusion.

Diffusion Pressure Deficit (DPD): Diffusion pressure deficit is the pressure that opposes the diffusion of water from higher chemical potential to lower chemical potential. It can be represented by a relation DPD = OP-wall pressure (turgor pressure).

Dioecious: Male and female reproductive organs in separate flowers on the different plant. Some wild American grapes are dioecious, with male plants unable to produce fruit.

Disbudding: The process in which remove surplus buds,flowers, or shoots from a plant to improve its shape, quality, yield, *etc.*

Disease Incidence: It is a rate or range of occurrence or influence number of new cases per population at risk in a given time period.

Disease Severity: The measure of the damage done by a disease.

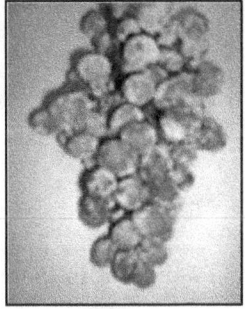

Disease/Disorder: Abnormal growth and dysfunction or some disturbance in normal functioning of a plant. Abiotic disorder-caused by environmental conditions such as frost, hail, and chemical burn.

Biotic disorder- caused by fungi, bacteria, and viruses.

Disinfectant: Destroy or inhibits the growth of disease-carrying microorganisms and harmful organisms by using physical or chemical agent/method.

Disinfest: To destroy infectious agent/microorganism that not still created any disease or disorder or harmness to the surface of object like soil, tools, plant.

Dispersal: The process of spreading of pathogen or microorganism within a given area or over the earth.

Distal: This is the point of attachment at the opposite end.

Diurnal: A daily or each day or during the day cycle that is completed every 24 hours or course of a calendar of the day.

e.g., some flowers opened during the day and closed at night.

Diuron(3-(3,4-Dichlorophenyl)-1,1-Dimethyurea): It is selective herbicide used to kill broad-leaved and grassy weeds. It is quite effective in vineyard, particularly when applied to a bare and clean soil before germination of weed seeds. At higher concentration, its effect persists for six months.

DNA: Deoxyribonucleic acid is a long polymer made by building blocks called as nucleotides. It is double helix structure encoded with genetic information.

DNA Fingerprinting: It is a technique used in the laboratory to identify an individual at the molecular level or detection of an individual by the genetic markers, which are present on chromosomes.

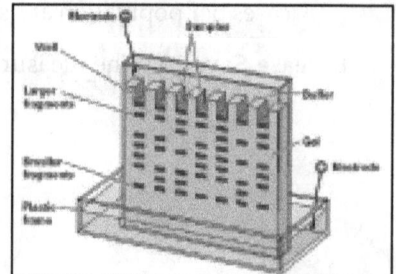

Domestic Indian Wine: The wine produced from and bottled in India for the domestic market.

Dormancy: A period at which temporary slowing or cessation or stopped plant growth activity especially in winter or dry seasons. Dormancy tends to be closely associated with environmental conditions.

http://en.wikipedia.org/wiki/Dormancy

Dormant Pruning: Pruning during the dormant season.

Dorsal: The upper surface of the leaf.

Downpour: A heavy continuous fall of rain.

http://www.thefreedictionary.com/downpour

Downy Mildew: This a very serious disease of grapevine. White or grey 'bloom' on leaves and stems caused by the production of sporangiophores and sporangia by members of the Peronosporales (downy mildew fungi). It is the most devastating disease of grapes in the tropical regions where there is continuous rainfall, and the temperatures remain above 10°C during the susceptible growth stages. Fungus causes it. It reduces vine productivity and fruit quality. All commercial grape varieties (*Vitis vinifera*) are highly susceptible. White downy growth on the lower surface of leaves with a corresponding yellow patch on the upper surface. Petioles, tendrils, young shoots, inflorescence and berries also attacked. In severe infections, infected tissues turn brown, dry up and drop. Infection at the time of flowering can cause 100 per cent yield damage. The fungus primarily survives in the infected stem as dormant mycelium.

Dried on the Vine: The process in which grapes bunches dry naturally on the vine plant to result in raisins. Naturally dried raisin consists different flavour and test than the artificial drying.

Drip Irrigation: It is a method of irrigation where water is provided to the grapevine drip by drip in precise amounts by a system of pipes and metered valves. Modern vineyards equipped with sensor technology may have their irrigation pattern computerized with the amount of water being adjusted depending on the data received from the soil sensors.

http://en.wikipedia.org/wiki/Glossary_of_viticulture_terms

Drizzle: Gently and steadily raining drops of precipitation consisting of numerous, tiny water droplets less than 0.5 mm in diameter.

http://dictionary.reference.com/browse/drizzles

Drought: A continuous period of dry weather due to lack of water/Moisture/rain, it can have a substantial impact on the ecosystem and agriculture of the affected region. It 's hard to decide the time of it started and its end. Three types of drought -Meteorological, Agricultural, Hydrological drought.

Dry Bulb Thermometer: Air temperature is measuring instrument.

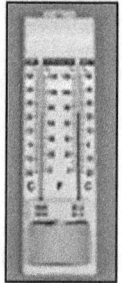

Dry Spot: Single berries in a bunch develop sunken, brown and dry spots of irregular shape before softening. Such berries do not ripen. No pathogen is associated with the dried spots of berries.

Dwarfing Rootstock: A rootstock that limits the size/height of the plant that is grafted onto it.

Dwarfness: The genetically controlled reduction in plant height.

E

Ecology: The branch of biology that deals with the study of relations and interactions between organisms and their environment.

http://dictionary.reference.com

Economic Injury Level: Population density at which the cost to control the pest equals the amount of damage; it inflicts, number of insects per unit area or per sampling unit.

http://www.cals.ncsu.edu/course/ent425/tutorial/economics.html

Eco-physiological Disorder: This type of disorders is caused by physiological as well as environmental factors. Eco-physiological disorders are those caused by the physiological disturbances under the influence of unfavorable environment. Such disorders appear in one year in a vineyard, but do not appear in the subsequent years.

Edaphology: Edaphology is the science that deals with the influence of soil and other media on the growth of plants.

http://agsci.psu.edu/elearning/course-samples/TURF_434/Ln_1/L1_4.htm

Effective Precipitation/Rainfall: Effective precipitation or rainfall is that part of the total precipitation on the cropped area, during the time period, which is available for plant development and transpiration. The portion of the total precipitation which becomes available to the growth of the grape plant.

Electrodynomorphic Soils: These are the soils that have been influenced by other environmental factors rather than the parent material.

Electromagnetic Radiation: Waves are produced by the motion of electrically charged particles; these waves called as electromagnetic radiation. The behavior of electromagnetic radiation depends on its frequency. All kinds of electromagnetic radiation are released from the Sun; our atmosphere stops some kinds from getting to us.

Electrophoresis: A technique used for separation and identification of molecules such as deoxyribonucleic acid or proteins using an electric current.

Emasculation: Removal of stamens from bisexual flowers before the anthers is mature to avoid self-pollination.

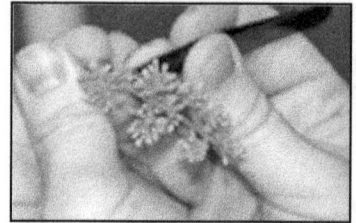

Embryo Rescue: It is processed promote the development of an immature or weak embryo into a viable plant by using *in vitro* techniques. It helps for producing interspecific and intergeneric hybrids.

http://naldc.nal.usda.gov/download/42085/PDF

Embryonic Bunches: During the annual cycle of the grapevine number of green berries arranged into bunches on a branch of the vine. A bunches will eventually bloom during the flowering period and, if fertilized, will develop into fully formed grape clusters. It's indicated potential crop yields.

Endemic Disease: A disease that regularly or always found in a certain class variety or animal or plant at particular area/location.

Endodynomorphic Soil: The soils with properties that have been influenced mainly by parent material.

http://forest.ap.nic.in/GlosTech-E.htm

Endogenous: Originating or developing from internally or within plant or organism.

Endoparasite: A various parasite that lives in the internal organs or inside of its host, feed on their host. Disrupting their hosts' nutrient absorption, causing weakness and disease.

Enology: It is branch of science which deals with study of all aspects of wine and winemaking.

Entomology: It is a branch of science which deals with the study of insects. vineyard affecting by the sucking insect pests namely thrips, hoppers, mealy bugs *etc.*, and beetle pests like stem borer, stem girdler, flea beetles, chafer beetles, shot hole borer and several lepidopteran, mites, nematodes and vertebrate pests.

Entomopathogenic Nematode: Entomo-pathogenic nematodes are soft-bodied, non-segmented roundworms, lacking appendages, predaceous or parasitic in nature occur naturally in soil environments.

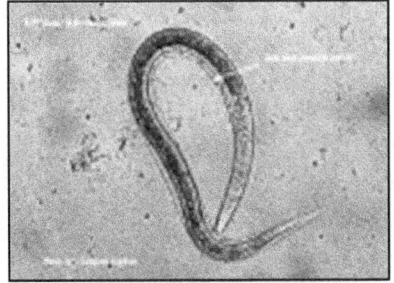

Entrepreneur: A person or enterprise that is willing to initiate a marketing operation, bringing together the necessary resources and taking the risk of success or failure.

Crop Post-Harvest: Science and Technology, Crop Post-Harvest: Principles and.edited by Peter Golob, Graham Farrell, John E. Orchar; page 514.

Environment: The total of all surroundings or external conditions especially those in which people live or work, an example Ecology in which interactions among organisms and their environment.

Enzyme: It is complex proteins that cause a specific chemical change in all parts of the living body. Pectinases structure changes during ripening of berries after that grapes become softer. Grape berries contain oxidative enzymes (tyrosinase).

Enzyme-Linked Immunosorbent Assay: These are also used as analytical tools in biochemical research for the detection and quantification of specific antigens or antibodies in a given sample.

http:///jid/journal/v133/n9/full/jid2013287a.html

Enzyme-Linked Immunosorbent Assay (ELISA): Enzyme-Linked Immunosorbent Assay is analytical biochemistry assay method, used to determine particular protein or amino acids are present in a sample and if so, how much it is. It also applied to detection of antibodies in blood.

Epidemic: The disease spreads very quickly and at the same time affect a large number of individuals within a population community or region.

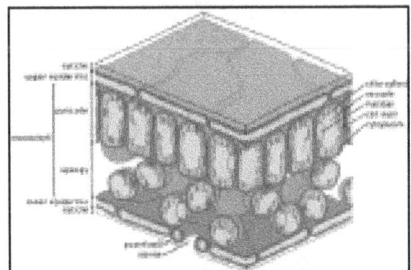

Epidermal Cell: Epidermal cells secrete the waxy hydrophobic substance cut in that polymerizes on the surface, forming a barrier to water evaporation.

http://www.biologyreference.com/Co-Dn/Differentiation-in-Plants.html

Epidermis: Thickened outermost cellular walls with multifunctional tissue; which cover and protect a whole plant body part like leaves, flowers, roots and stems. Plant have three type epidermal cell- pavement cells, guard cells and their subsidiary cells that surround the stomata and trichomes.

Epiphytic: Often called "air plants," plant that derives moisture and nutrients from the atmosphere and rain; develops on another plant but not parasitic in nature.

Equator: It is zero-degree latitude on the earth's surface. It is equal distance from the North and South Poles and divides the Northern Hemisphere from the Southern.

Equivalent Acidity: The number of parts by weight of calcium carbonate (as $CaCO_3$) required neutralizing the acidity resulting from the use of 100 parts by weight of the fertilizer.

Equivalent Basicity: Equivalent basicity means the number of parts by weight of calcium carbonate (as $CaCO_3$) that corresponds in acid neutralizing capacity to 100 parts by weight of the fertilizer.

Definitional Glossary of Agricultural Terms: By Dinesh Kumar; Y. S. Shivay; vol 1 page 92.

Eradicant: Destroy completely or root out pest, pathogen or disease creating organism from area or host by using chemical, biological control method, destruction of the diseased plants, and eradication of alternate host plants, pruning, disinfection and heat actions.

Eradication: Removal of pest or diseases by a physical method.

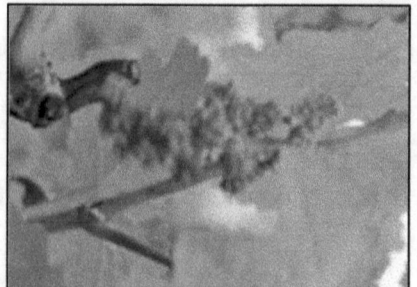

Erosion: The process by which soil or rock are removed due to the force of gravity from one area to other by exogenic processes such as wind or water flow or action of the sea.

Espalier: It is the technique of training for controlling the growth of grapevine, especially for shoot trained into flat, two-dimensional forms to grow in one plane. The vertical canopy of vine in rows.

Esters: Condensation of an acid and an alcohol formation of esters, general formula RCOOR, which are responsible for the pleasant fermentation aroma of wines. Many naturally occurring fats and essential oils are esters of fatty acids.

Ethanoic Acid: Also called as acetic acid, glacial acetic acid. Chemical formula- CH_3COOH, vinegar is a liquid consisting mainly of acetic acid (CH_3COOH) and water. There are several different qualities of red wine vinegar.

Ethanol: It's structural formula CH_3CH_2OH. Also known as ethyl alcohol. This is primary alcohol in wine and most other alcoholic beverages. Ethanol is intoxicating agent occurs in wine and beer.

Eutypa Dieback: The fungus Eutypa lata causes a Eutypa Dieback Destructive woody tissue disease of grapes. It is found in commercial grape production, viable in the field trunks of infected living vines and dead grape wood. Most frequently in vineyards established for 10 or more years.

Evaporation: The physical process by which a liquid, such as water is transformed into a gaseous state, such as water vapor. It is the opposite physical process of condensation.

http://www.indiaweather.in/Glossaries/gloss_e.aspx

Evapotranspiration: Transpiration+Evaporation=Evapotranspiration

Hydrologic cycle in which transport of water into the atmosphere from Earth surface including soil (evaporation) and vegetation (transpiration-leaves, stem, flowers, and/or roots) depends upon climatic condition.

Exclusion: A phenomenon in which existing infection prevents from an infected area or plant material by excluding other.

Excretion: Excretion is process in which elimination of nitrogenous waste product from living body, blood, tissues or organs through excretory system.

Exocarp: Exocarp is the botanical term, the outer layer of the pericarp (or fruit). It forms the skin of a grape or peach; skin contain oil glands and pigments.

Expeller: The machine used to crush grape oil seeds to yield oil.

Exposure Cluster to Sun: In the absence of adequate foliage and improper orientation of the shoot, even in the presence of adequate foliage, bunches are exposed to the sun during ripening. In such bunches, berries get tanned and developed amber brown colour on the upper portion, is called sunburn berries.

Extra Dry: It is wine with small amount residual sugar. These wines are mostly acidic in nature.

Extracellular: Located or occurring outside a cell or cells.

http://www.thefreedictionary.com/intracellular

Extract: It is processed for isolation or separation of substance or desire product from raw material by using solvent. For improving wine quality flavor, tannin, colour and other phenolic compounds extracted from grapes.

Exudate: A substance that oozes out.

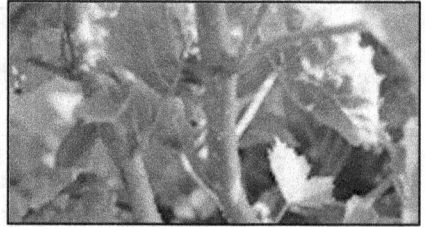

F

Farm Efficacy Measurement: The ratio of input to output means efficacy. Farm efficiency help to evaluate profitable farming business with cost and return.

Farm Management: Farm management is a decision-making science. It helps to decide about the basic course of action of the farming business.

http://beta.krishiworld.com/html/farm_management1.html

Farm Map: The farm is carefully mapped out giving its silent features, like soil type, soil fertility and rotation flow. It also helps analyzation of previous crop history.

Farm Visit: Farm visit constitute the direct contact with farmer by scientist or extension worker. During this visit discussion based on problem of farmer or organizational purpose can be discussed.

Farm Yard Manure: It consists of a mixture of animal dung, poultry manure, wood ashes, plant residue, straw, stalk and green manure crop. It is most commonly used organic manure in India.

Farmer: A person who operates a farm.

Farmer's Market: A place where farmers or producers sell their products directly to the consumer.

Fasciation: Abnormal flattening of stems due to injury or infection that results in failure of the lateral branches to separate from the main stem. Such phenomenon occurs in the stem, root, fruit, or flower head is called as Fascination.

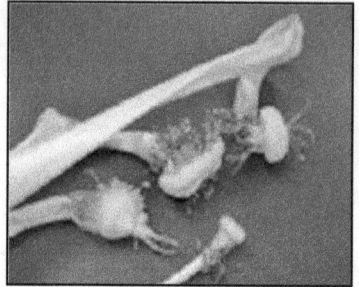

Fatty Acids: Fatty acids are merely carboxylic acids with long hydrocarbon chains vary from 12-18 carbon. A grape contains following acids-tridecanoic acid, myristoleic acid and butyric or propionic acids. Fatty acids also help to activate fermentation in grapes.

Feeder Roots: A complex system of smaller roots grows outward and predominantly upward from the system of framework roots. Feeder roots absorb water and minerals from the soil. A fine, short, non-woody tip looks mat-like structure.

Femur (Pl., Femora): It is largest, a thickest segment of the insect leg; attached to the body through trochanter, coxa, and distally to the tibia.

Fermentation: It is a process in which sugars gets converted into alcohol.

Fertigation/Nutrigation: It is a method of application of fertilizers through irrigation, particularly by drip irrigation system. All the forms of fertilizers cannot be used in fertigation as the insoluble salts can choke the irrigation system by clogging the emitters. Choice of the form depends upon the soluble salts in irrigation waters.

Fertility Index: Fertility index is defined as the relative sufficiency expressed as a percentage of the amount of nutrient adequate for optimum yield. It is related to soil test value and crop yield.

Fertilization: 1. Application of external plant nutrients to supplement the nutrients naturally occurring in the soil to maintain an optimum supply for plant growth, Nutrient may be organic fertilizer, synthetic fertilizer. Sixteen are essential plant nutrients. Different kind of fertilizer used for different situation. 2. In the process of fusion of gametes to produce a new organism.

Fertilizer, Inorganic: A fertilizer material in which carbon is not an essential component of its basic chemical structure. Urea is often considered an inorganic fertilizer.

Soil and Environmental Science Dictionary; edited by E.G., Gregorich, L. W. Turchenek, M.R. Carter - 2001; page no-132.

Fertilizer, Organic: A fertilizer material in which carbon is an essential component; it derived from green manure, post harvesting plant material, harvested vermicompost, peat, seaweed algae, humic acid called as organic fertilizer.

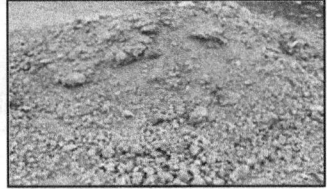

Fertilizer, Starter: A relatively small application of fertilizer applied with or near the grape plant for accelerating the early growth of the plant.

Fertilizer, Suspension: A fertilizer containing miscible and immiscible plant nutrient.

Fertilizers: Fertilizer is inorganic materials of a concentrated nature; they applied for increase supply of one or more of the essential nutrients *e.g.*, potash; phosphorus; nitrogen.

Field Experiments: Experiments conducted in vineyard to determine the type of diseases, soil fertility, water content, plant development, *etc.*

Field Grafting: It is an old technique for changing the grapes fruit varieties at a high level in the field without the expense of replanting and lost cropping years.

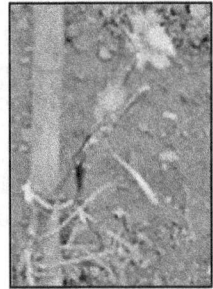

Field Heat: It is the heat accumulated by the plant tissue. It is also called as heat units.

Filament: The stalk of a stamen that bears and holds the pollen sacs. Filaments exist to carry nutrients to the anther.

Filtering: It is a technique used to remove impurities or suspended particle from grapes juice or wine to obtained desired product. Filtration is a more efficient method for the separation of mixtures. There are two reasons to filter wine: clarification and microbiological stabilization.

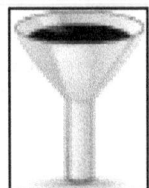

Fine Wine Lees: A wine Lee is a residue that forms at the bottom of recipient containing wine, after fermentation, during storage or after treatment.

Fining: 1-The process of specifically removes polymerized tannins, colouring phenols and proteins from wine, ensuring its future clarity. For ex - 1-Bentonite is a special fining agent added to remove protein in white wines.

2-Fining agents for red wines are animal proteins such as egg whites, gelatin and milk.

Fixed Acidity: Fixed acidity = total acidity – volatile acidity

Acidity of wine can be calculated by using titration method (volatile) or thin-layer chromatography(Non-volatile)

Fixed Sulfur: Major types of wines contain sulfur dioxide molecule depending upon types of wine in various forms, collectively known as sulfites. Sweet wine contains more sulfur per cent because it binded with sugar and acid a high proportion. The bound sulfites are those that have reacted (both reversibly and irreversibly) with other molecules within the wine medium.

Flavour Development: Monoterpenes are responsible for the creation of distinctive flavor in grape. Desirable flavour components accumulate later in ripening after the sugar increase slows down.

Flea Beetle: The flea beetle is known to feed on the buds and leaves leading to an adverse effect on the growth and production of grapes in the field. It is an insect grapevine pest that affects immediately after pruning until seven leaf stage and severely affect during bud sprouting stage. Newly sprouting buds are eaten fully. Defoliated leaves with characteristic slit cuts.

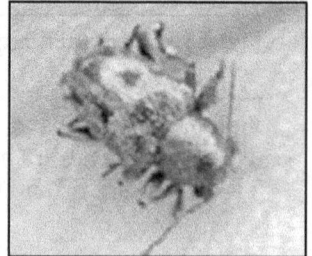

Flocculation: To aggregate or clump to gather individual soil particles, especially fine clay and humus, into small clumps or granules that usually settle out of suspension quickly.

Floral Biology: It is a detailed study of the flower types, time and duration of flowering, dehiscence if anthers, pollen germination, pollen tube growth and the receptivity of sigma are primarily important for implementing any hybridization program effectively and successfully.

Floral Differentiation: It is changes or advancement of floral stages since its initiation.

Flower Cap: It is a covering portion or part that comes off during flowering.

Flower Cluster Initiation: The period at which flower cluster structure begin to form on the primary stage shoot is called flower cluster initiation.

Flower of Grape: Grapevine flowers are born in a cluster (or bunch).

Fog: Formation of fog means, when water vapor condenses into tiny water droplets or ice in the air or near to earth surface that reduce visibility. Fog can form at a relative humidity near 100 per cent.

Foliar Diagnosis: Diagnostic technique based on analyzing plant tissue for the total content of nitrogen, phosphorus, potassium, boron, *etc.*

Dictionary of Soils and Fertilizers, By L. L. Somani; Vol 4, part 2 page-505.

Foliar Feeding: Technique of feeding plants by applying liquid essential nutrient directly through their leaves, foliar feeding be done in morning or evening.

http://www.netlibrary.net/articles/Foliar_feed

Foliar Phylloxera: It is a most destructive tiny pest; that causes galls or rigid swelling on the outer surface of leaves and roots of grapevines.

Forecast: To calculate or predict quantitative data about the future event or condition on the basis of the previous state. *e.g.*, in weather forecasting prediction made by collecting quantitative data about the current state of the atmosphere along with the skill and experience of a meteorologist.

Fortification: The process of addition of pure alcohol into wine during or after fermentation for the purpose of enrichment.

Foundation Pruning: It is known as back pruning. Healthy new canes must be produced every year to maintain annual production of fruit.

Foundation Stock: It is a true to type material free from viruses and used for further propagation.

Frass: It is a solid excreta of insects. Insect Frass comes from plants.

Free Run Juice/Nobel Juice: The juice that flows freely from freshly picked grapes skin that have not been pressed.

Free Sulfur: The free sulfites/sulphur are those available to react and thus exhibit both germicidal and antioxidant properties.

http://www.mdpi.com/1424-8220/12/8/10759/htm

French-American Hybrids: Hybrids of European (*Vitis vinifera*) and American grapes. This variety used for the production of good quality sparkling wines, intense red wine, fruity, mildly intense white wine.

e.g. Villard Blanc, Verdelet, Seyval Blanc, Ravat (Vignoles), Landot, Cayuga (White), Baco Noir

---

Fresh Water: Freshwater is defined as having a low salt concentration — usually less than 1 per cent. 3 per cent of earth's water as fresh water. Fresh water sources are ponds, lakes, stream, river.

http://www.ucmp.berkeley.edu/glossary/gloss5/biome/aquatic.html

---

Frost: Frost forms through sublimation, a deposit of tiny, white ice crystals on a surface from the humid air; when water vapor in the air condenses at a temperature of freezing. Frost may harm crops or decrease future crop yields.

Fruit Bud Formation: A bud that produces flowers and then the fruit will form. Thompson Seedless flower buds usually have one or two inflorescence.

Fruit Growth: After successful pollination and fertilization of ovules within flower berry formation takes place. only 20-30 per cent of flower on cluster developed into mature berries.

Fruit Pruning or Forward Pruning: October pruning is called fruit pruning or forward pruning. The number of buds retained on a cane at forward pruning depends on variety and cane thickness.

Fruit Set: A development and maturation of a quiescent ovary to a rapidly growing young fruit, which is an important process in the sexual reproduction of flowering plants.

http://dirtywordsgarden.com/vocab/

Fruit Wine: Fermented alcoholic drink made from ingredients like fruit (not include grapes) rhubarb, elderberries, bananas, coconuts and cranberries; and additional flavour taken from herbs, flower and sometimes sweetness increased by adding honey or sugar externally.

Fruitful Bud: A bud that has a rudiment bunch.

Fruiting Body: Spore producing structures, species to species size, shape and colouration changes which aid in identification of the particular fungus.

Fruiting Cane: A cane is having fruitful bud.

Fruiting Wood: Canes or spurs that are selected for their size and quality and cut back to bear the current year's crop.

http://www.mdtgrow.com/pruningterms.html

Fruiting Zone: It is the portion or buds on the shoots those are fruitful.

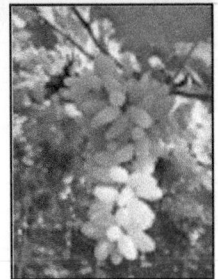

Fumigant: It is vaporized chemical; a poisonous substance used to kill insects, pest. An example - a mixture of CO_2+SO_2 used to control black widow spider in grapes. Methyl bromide used to control arthropod pests in grapes.

Fungicide: Fungicides are a chemical compound, or biological organisms prevent the growth of fungi and their spores.

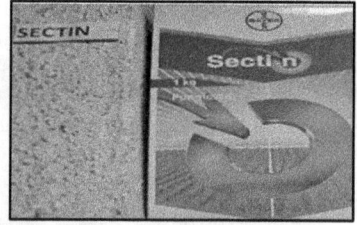

For Example-Bordeaux Mixture - (copper sulfate and hydrated lime in water) used to control downy mildew and powdery mildew of grape.

http://www.hindawi.com/journals/jb/2012/135479/

Fungus/Fungi: A single-celled or multicellular organism that's usually walled, threadlike cells; absorb nutrients which depend on another organism for nourishment. Fungus is different from plants and animals.

Gall/Hypertrophies: Gall is an abnormal vegetative swelling response by plants tissue on stems, leaves, flowers and roots; injury caused by microbes, fungus, mistletoe, mites, nematodes, virus, and insects.

Generation, Insect: Insect generation is the time needed to complete one generation. Insects of short generation time have higher rates of increase and develop resistance to insecticides more quickly than insects of comparatively longer generation time.

http://entnemdept.ifas.ufl.edu/walker/ufbir/chapters/ chapter_06.shtml

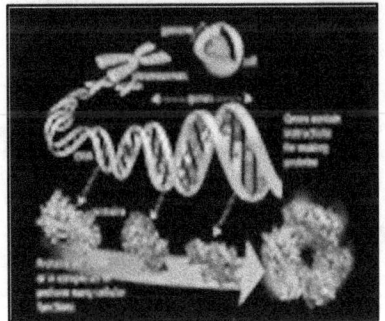

Genome: The genome is the heritable material of an organism. Genomes include together gene and the DNA/RNA non-coding order.

Genotype: Genotype means the entire set of genes in a plant cell.

OR

The genetic constitution of an individual or organism distinguished from the physical appearance.

e.g., The gene responsible for plant height, colour, size, fruit.

Genus: The taxonomic division of classification below family and above species. *Vitis* is a genus of grapevine plant.

Geographical Indications: GI are signs used on goods that have a specific geographical origin and possess qualities or a reputation that are due to that place of origin. Agricultural products typically have specific qualities like colour, test, flavor and aroma that derive from their place of production and are influenced factors, such as climate and soil.

Geographical Information System: A geographic information system (GIS) is a computer system designed to capture, store, manipulate, analyze, manage, and present all types of geographical data.

Geosmin: A powerful sesquiterpene compound with earthy smell found in grape wine. Geosmin will be stable during alcoholic fermentation and storage, particularly at optimum temperature for aging wine (15°C).

Geosphere: The geosphere is a rigid outermost shell of a rocky planet of the earth, including atmosphere, hydrosphere and biosphere.

Germination: It is process by which plants, fungus and bacteria emerge from seeds and spores and begin growth. As seeds imbibe water, hydrated enzymes become active and the seed increase its metabolic activities to produce energy for the growth process.

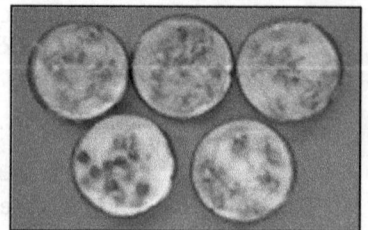

Germplasm: Germplasm contains the information for a species' genetic makeup, a valuable natural resource of plant diversity. Germplasm help to researcher and grower to develop new high-yielding, high-quality varieties make management easier. A germplasm is a collection of genetic resources for a plant/organism.

Gibberellins, Gibberellic Acid (GA): Gibberellins are diterpenes. It shows many functions each depending on the type of gibberellin present as well as the species of plant. The function of gibberellin-breaks seed dormancy, enzyme production, stem elongation, delay senescence in leaves, cell division and elongation.

Girdling: Remove an outer ring of bark, trunk, or woody plant part to prevents the translocation of carbohydrates to the root system. Girdling is often used only when necessary, in table grapes production. Girding is also known as ringing or cincturing. It consists of removing a narrow ring of bark around the cane, arm or the trunk. The effect of girdling on increasing the fruit set in grapes is an accidental finding.

Glabrous: Plant without hair.

Glaucous: Covered with a whitish waxy substance or pale grey or bluish-green colour appearance; it is also known as bloom.

Global Positioning Systems (Gps): GPS is a satellite-based navigation system for mapping vineyard environmental variability. GPS works in any weather conditions, anywhere in the world, 24 hours a day.

Glyphosate (N-Phosphonomethylglycine): It is also non-selective, broad spectrum herbicide. It is very effective on deep-rooted perennial, biennial species of plant.

http://umaine.edu/blueberries/factsheets/weeds/237-glyphosate-for-weed-control-in-wild-blueberries/

Grading: It is mechanical or manual process of separation resin or berries. Grading of grapes on the basis of the size of the bunch, colour, texture, flavor, and aroma is also done. Mechanical grading now used for separation of rasin. Mostly size gardening used in India.

Graft Compatibility: Close genetic (taxonomic) relationship of scion and rootstock to join and grow.

Graft, Grafting: Grafting is horticultural technique in which stock, and scion plants must be placed in contact with each other taking apart from one plant and making it grow on another plant.

Grape: Grape is one of the most commercially important crops of world and is a fairly good source of minerals like calcium, phosphorus, iron, vitamins-like B_1 and B_2.

Grape Antioxidant: Grape or grape wine contains chemical Phyto constituent such as proanthocyanidins, polyphenols, flavonoids, and anthocyanins, these all shows high antioxidant properties.

Grafted Vine Shift to G: A grafted vine consists of two parts, the scion variety, which produce the fruit and the rootstock variety that provides the root system and lower part of the trunk called as stock.

http://www.extension.org

Grape Arm: Small branches of the trunk from which canes or spurs grow.

Grape Embryo: The embryo grows within the developing seed while the entire ovary grows to become the grape berry itself with seeds contained within.

http://www.hort.cornell.edu/reisch/grapegenetics/breeding/crossing1.html

Grape Guard: It is a sulphur dioxide releasing pad. It is used for packing grapes for long distance markets to protect grapes from developing fungal infections.

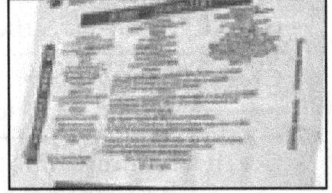

Grape Harvesting: Harvesting of table grapes when the flavor is right; harvesting grapes when they reach the appropriate sugar and acid content.

Grape Leaf Roller: This is a serious pest in South India. They are found to be active from August-November. The adult has Yellowish-green caterpillars. Leaves show characteristic rolling symptoms. In case of severe infestation, complete defoliation is observed.

Grape Macro Nutrient: Nitrogen; Phosphates; Potassium; Calcium; Magnesium; Sulfur

Grape Micro Elements: Iron; Boron; Manganese; Zink; Copper

Grape Molasses: It is thick sweet syrup that is made from the grape juice that has been boiled down until it forms a syrup with a concentrated fruity grape flavor.

Grape Nitrogen Deficiency: Nitrogen deficiency means entire leaf area turns pale green and then yellow. Reduced size and thickness of leaf, weak vines and poor crop growth.

Grape Pomace: The skins, stalks, and pips (seeds) that remain after making wine. This is also called pomace.

http://www.the-gift-of-wine.com

Grape Sorting Table: Wine grape sorting table used in the process of winemaking used to sort out unwanted MOG. (Material Other than Grapes).

http://www.thevintnervault.com

Grape Stalk Crusher: It used for reducing your grape stalk volume by upwards of 80 per cent.

http://www.thevintnervault.com

Grape Vine Cold Hardness: It is highly dynamic condition, influenced by environmental and growing condition, and varying among grapevine varieties and tissue and over time.

http://symposium.surry.edu/schedule/category/Viticulture/past/

Grapes Anther: It is the pollen containing part of the stamen.

Grapes Juice: Liquid extract of the fruit grapes prepared by mechanically squeezing, crushing, blending or pressing of grapes. Grape Juice always - better than wine.

Gravel: Coarse soil particle is having 0.2 to 2 cm in diameter.

Green Harvest: The removing of green (unripe) grapes from bunches to increase growth and the yield of desired quality grapes.

Green Manure Crop: Green manure means planting a crop usually belonging to leguminous family and inserted into the soil to enrich its fertility after adequate growth.

Greenhouse Effect: The overall warming of the earth's lower atmosphere primarily due to carbon dioxide and water vapor which permit the sun's rays to heat the earth, but then restrict some heat energy from escaping back into space.

http://dictionary.babylon.com/greenhouse_effect/

Groundwater: Water below the surface of the ground at a pressure equal to or greater than atmospheric.

Soils: Genesis and Geomorphology - By Randall J. Schaetzl, Sharon Anderson, page No- 760.

Grow Tube: Grow tube usually around, plastic tube that fits around young vines designed to protect young trees. The advantage of grow tube - Faster establishment of balanced, healthy vines, Protection from herbicide spray, Reduced vine training costs, Protection from animal predation.

Growing Season: A growing season is the period of the year when crops grow actively. Growing season determined by climate and elevation.

Growth and Development: Growth is physical process refers to increase in berry size and development refers to the biochemical changes in which berries are ripening starts.

Growth Promoter: A natural or synthetic substance that improve the overall health growth and development of plants. It also helps in improving the crop, quality and productivity significantly. *e.g.*, Gibberellic acid, Indole butyric acid and Indole acetic acid.

Growth Regulator: Growth regulator is a natural or synthetic substance or mixture of substances increase or slows down the rate growth or maturation through their physical or chemical action.

e.g., Ethylene is natural growth regulator.

Growth Retardant: A natural or synthetic compound that reduces or stops plant growth. It is an effective strategy for controlling plant height by using the proper growth retard technique.

OR

Chemical that slow cell division and cell elongation in shoot tissues and regulate plant height physiologically without formative effect.

Guard Cells: Guard cell is crescent-shaped cells that surround a pore (stoma) in the epidermis. Stoma opening and closing regulate gas exchange and water loss. Guard cell also has chloroplast pigment and conduct photosynthetic effect.

Gum: Gum is gelatinous, sugary substance and adhesive substances that is synthesized and secreted from the disintegration of internal plant tissues or decomposition of plant cellulose. Some types of gum dissolve in water or some produce mucilage by absorption of large quantities of water.

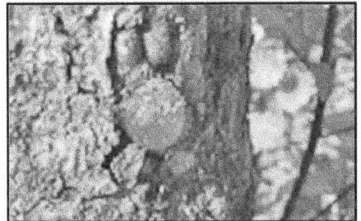

Gummosis: The copious production and exudation of gum by a diseased or damaged tree especially as a symptom of a disease of fruit trees. Oozing out resinous substance from the damaged part of the tree.

Guttation: The extracellular secretion of exudation drops of xylem sap on the margin of leaves at night called as guttation. Such drops contained different compounds, like sugar, mineral nutrients, and potassium.

Gypsum: A hydrated calcium sulphate ($CaSO_4.2H_2O$). The commercial material contains varying amount of impurities and is used as a soil amendment.

http://www.bis.org.in/sf/fad/FAD72538.pdf

Hail: Hail is showery precipitation usually falls during thunderstorms in the form of spherical or irregular pellets of ice larger than 5 millimeters (0.2 inches) in diameter. Individual lumps are called hailstones.

Hailstorm: Hailstorm is occasionally experienced in the peninsular India during the shoot growth period after April pruning. A hailstorm occurs during the fruiting period, damaging the cluster and reducing the yields of fruit plant.

Halophite: A salt-loving organism that tolerates saline conditions.

Hardpan: A hard, usually clay-rich layer of soil lying at or just below the ground surface in which soil particle cemented together by silica, sesquioxides, or calcium carbonate. Hardpan does not dissolve in water, also called caliche.

Hard-Wood Cuttings: Hardwood cuttings are taken from deciduous trees and plants. (ones that lose their leaves in winter) Cuttings are made from the new shoots (canes) that grew the growing season that just ended. This technique is very useful for propagating grapes trees.

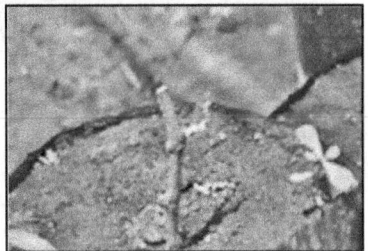

Hardy: Characteristic of plants that are tolerant of adverse climatic conditions, usually used with reference to cold tolerance.

http://eviticulture.org/glossary-of-grape-terms/

Harrow: The harrow is used for lighter operations like to tillage in heavy soil. Harrow are equipped with discs or spring tines or spikes.

Harvest Index: Harvest index defines the criteria to judge the optimum stage of maturity of grapes for the harvest. It is broadly based on the physical appearance of the bunch, climatic factors and biochemical composition of berries.

Harvestability: The gathering of a ripened grapes.

Head System: It is a least expensive grapevine training system. It is best suited for less vigorous varieties with very less degree of apical dominance, and for those in which the basal buds in a cane are fruitful.

http://www.nhb.gov.in/Horticulture per cent 20Crops/Grape/Grape1.htm

Heat Balance: Heat balance means, it is a comparison of the heat loss and heat gain from sun, atmosphere and earth.

Heat Index: The Heat Index is a measure of how hot it feels when a combination of air temperature and humidity takes place.

Heat Summation: It is a growing degree days required for the attainment of the any phenological stage

Hectare: A hectare is a unit of area equal to 10,000 square meters.
1 ha = 2.47 Acres

Hedging: Hedging is necessary on non-shoot-positioned training systems. When shoots have grown well beyond the top wire eliminating the shoot tip can stimulate the growth of lateral shoots from the nodes to keep vine growth appropriate for a given trellis system.

Heliotropism: The directional growth of a plant either vertical or equivalent to the sun's radiation called as heliotropism.

Hemic Material: Organic residue in organic soils is having an intermediate stage of decomposition.

Herbaceous: The herbaceous plant that does not form a persistent woody stem. It's stems are green and soft. A term often used to describe the flavors in wine or grapes that resemble leaf, or vegetative flavors.

Herbicide: Herbicide is chemical used to kill unwanted weeds or plant or vegetation.

Herbicide Injury: The herbicides like 2,4-D is very much dangerous to vine health and most of the time vine dies due to the sprinkles of this herbicide.

Heterozygosity: The species of grapes are quite heterozygous and seedling offspring's show wide genetic variability.

High Density Planting: High density means to increase the vine plant population per unit area for increasing the quality production of fruit crops; and best utilization of limited land and resources.

High-Performance Liquid Chromatography (HPLC): High-Performance Liquid Chromatography is a technique used for isolation or separation compounds from a mixture.

Histosols: A soil water regime is usually having adequate water during all of the growing periods of the year.

Hogshead: A unit of capacity used especially for measurement for alcoholic beverages.

http://www.memidex.com

Honeydew: Sweet excretory product secreted by caterpillars, mealybugs, soft scales when they feed on plant sap.

Hormone: Hormones are chemicals that regulate plant growth; they play a role in cell division, cell differentiation, fruit development, the formation of roots from cuttings, and in leaf fall (abscission).

Five general class of hormones are -Abscisic acid, Auxins, Cytokinin, Ethylene, Gibberellins

Host: A plant or animal depends upon another species (body surface or inside the body) for their growth and development. For example; the parasite obtains nutrition directly from the body of the host.

Hot Spot: It is presence of large amount of pest at particular pest

Humidity: Humidity means amount of water vapor in the air. Humidity depends on water vaporization, condensation and temperature. It regulates air temperature by absorbing thermal radiation both from the Sun and the earth.

Humus: Humus is a mixture of residue left after the partial decay of organic substances in and top of soil. A chemical property of the soil is more affected by humus than other sandy or silt particles. The alkalinity of a soil can be reduced slightly by the incorporation of humus.

Hybrid: A hybrid is a cross between two different plant varieties and produced new desirable variety. Hybrids technique is used to develop for disease resistance, size, flowering, colour, taste. Types of hybrids - double cross, single cross, triple cross hybrid.

Hybrid Grapes: A grape variety produced from parent vines of two different species by using contrast crossing technology – such as *Vitis vinifera* and *Vitis labrusca*.

Hydathode: A leaf gland that secretes water through stoma on the margin of the leaf.

Hydrogenic Soil: Soil developed under the influence of water standing within the profile for considerable periods, formed mainly in cold, humid regions.67

Definitional Glossary of Agricultural Terms: By Dinesh Kumar; Y. S. Shivay; vol 1 page-130

Hydrometer: A hydrometer is an instrument used to measure weight or gravity of a liquid in relation to the weight of water. Winemaker used a hydrometer to calculate the amount of sugar in must or wine, the progress of fermentation, percentage of alcohol. The typical hydrometer measures three things: specific gravity, potential alcohol and sugar.

Hydrosphere: A hydrosphere is the total amount of water at or near a surface of the earth. The source of the hydrosphere is ocean, lakes, river, well, aquifer.

Hygrometer: It is tool for measuring the amount of humidity in the atmosphere.

Hypha: A hypha is a long, branching threat like the filamentous structure of a fungus. Hyphae are produced on both upper and lower surface of leaf in powdery mildew disease.

Hyriic Soils: Soils that are periodically wet long enough to produce anaerobic conditions, thereby affecting the growth of plants.

http://www.fws.gov/habitatconservation/nwi/wetlands_mapping_training/module2/ DW4.html

I

Immune: Resistant to a particular infection.

Immunity: The ability of the body to resist or fight off infectious agents like virus or bacteria. Immunity may be natural or acquired. For *e.g.*, resveratrol from red grapes and pterostilbene from blueberries in the presence of vitamin D, increase the immune system of the organism.

In vitro: It is a biological process for conducting an experiment in a laboratory vessel or other controlled experimental environment (culture medium) outside the living organism.

In vivo: Medical test or experiment or procedure that is done on living organism by the different experimental condition.

Inceptisols: A layer of soil used in cultivation.

Indian Grape Processing Board: Indian grape processing board www.igpb.in is official website to provide knowledge, information and education about Indian Grape Processing board and its activities.

https://www.rankwise.net/www.igpb.in

Indian Sub-Tropical Climatic Region: All grape growing areas of Tamil Nadu, and the districts of Bangalore, Kolar and Mysore of Karnataka.

Indicator Plant: Indicator plant is a usually weedy plant that grows in some specific environment, allowing an assessment of soil and other conditions in a place by simple observation of vegetation.

http://translate.academic.ru/environment per cent 20indicator/en/ru/1

Indigenous Plants: Plant born, grown or produced naturally (native) for years in a particular region, without human intervention. They have adapted to the geography, hydrology, and climate of that region.

Indoleacetic Acid (IAA): Indoleacetic Acid is naturally-occurring, phytohormone of the auxin class; that stimulates cell elongation and cell division in root and stem part ($C_{10}H_9NO_2$). Often abbreviated as IAA.

Infection: It is the invasion of the grape tissues by microbial agents like bacteria, viruses, and parasites that are not normally present within the body. For Example -powdery mildew, black rot, downy mildew is disease infecting agent in grape.

Infest: Attack of insects or pathogen on the vine or its parts in large numbers, typically so as to cause damage or disease.

Infiltration Rate: It is velocity or speed at which water enters from the soil surface into the soil. It is useful for determining normal soil wetness condition prior to irrigation.

Inflorescence: Group or cluster arrangement of flowers on a central axis. It is described by the way of arrangement of flower on the peduncle, blooming order of flower, cluster group of the flower.

Infrared: An infrared ray is an electromagnetic radiation having wavelength above microwave but below visible light. Thermal radiation emitted by all hot objects near room temperature is infrared. Compositional analysis of grapes, juices, wines, and other alcoholic beverages done by Infrared Spectroscopy.
Near-infrared: 0.78–3µm
Long-wave infrared: 50–1000 µm.

Inoculate: To treat or prevent a disease introduce immunologically active material into the body (antigen or antibody)/To introduce a microorganism or infectious material into culture medium for suitable growth.

Inoculation: The process of introducing the culture of the microorganism into soils or culture media.

Inoculum/Inoculant: Source of the pathogen that cause disease infection in plants.

Insect Growth Regulator: Substance (chemical) that inhibits/interferes with the life cycle of an insect either altering growth hormones or altering the production of chitin.

Instar: Instar is a developmental phase of an insect larva between two periods of moult. Development rate always depends on temperature and humidity.

Integrated Agriculture: Integrated agriculture word to is used a more integrated approach to farming. It includes farm productivity, farm employment, income, environment and democratization.

Integrated Control: The control strategy that uses all appropriate methods and techniques (cultural, chemical, and biological) to control disease and pest populations to minimizing economic loss.

Integrated Pest Management: Integrated Pest Management is a strategy for selective use of agrochemicals, biological methods, genetic resistance, and appropriate management practices.

Intellectual Property Right(IPR): Intellectual property rights are legal rights, which result from intellectual activity in industrial, scientific, literary and artistic fields. Some important IPR types are - Patents, Trademarks, Copyrights, Related Rights, Geographical Indications, Industrial Designs, Trade Secrets Layout, Design for Integrated Circuits, Protection of New Plant Variety.

Inter Institutional Agreement: A legal agreement between the institute or university with other party to accomplish a particular task; which may be related to research and/or technology transfer and/or intellectual property management agreement for specific period of time.

Intercropping: Intercropping is alternative strategies applied in multiple cropping. The growing of two or more crops in proximity; as a result, two or more plants are managed at the same time. The four basic arrangements used for intercropping are row intercropping, mixed intercropping strip intercropping and relay intercropping.

Internal Soil Drainage: The underground movement of water through the soil is called as internal soil drainage.

Internode: Internode is the section of stem between two nodes. Internode part of the stem is made by some vascular vessels which carry water, hormones, and food from node to node.

Interveinal: In between the veins. Some nutrients deficiencies will occur interveinal in leaves.

Invention: Invention means unique process, composition, device or any product has shown technological advance than the existing one.

Inventive Step: An invention is considered to include an inventive step if it is not obvious to a skilled person in the light of the state of the art.

Inventor: Individual, who is the first to think of or make something as defined by patent law,; devices, some new process, appliance, machine or article. It is often considered to be a creator who materially and substantially contributes to the conception of a patentable Invention.

Inventor Disclosure: It is a confidential document written by a scientist or engineer for use by an institute's patent department or external patent attorney to evaluate and protect the process for Intellectual Property.

Iron: Iron is necessary micronutrient for the synthesis of chlorophyll. Iron deficiencies are associated with high pH soil conditions (soils with free lime). Yellow colouration of leaf with green veins is symptom of Iron deficiency.

Irrigation: The artificial supply of water to land, to maintain or increase yields of food crops. *E.g.,* Drip irrigation is a newer method mostly used by the grape grower to supply water directly to the base of each plant. Now a day various types of irrigation techniques are available in the market for distribution of water within the field. *e.g.,* sprinkler, dripper, surface irrigation, manual irrigation using buckets.

Irrigation Application Efficiency: Percentage of irrigation water applied to an area that is stored in the soil for crop use.

Jam: Jams are usually made from berries pulp and juice of grapefruit heated with water and sugar to activate its pectin. It appears like semi-jellied texture.

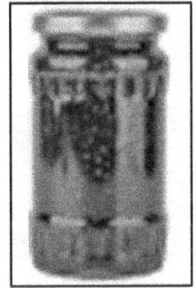

Jassid/Leaf Hoppers: It is a destructive pest of grapevines in North India and certain pockets in South India. 1) They are yellowish-green coloured wedge-shaped insect measuring 3 mm in length. 2) On leaves due to sucking of sap from them finally brown, dries up and drop off.

Jelly-Grape: Grape jelly is a semisolid gelatin mass of fruit extract. Fruit product usually made with gelatin or pectin by concentrating the sugar and the juice of grapefruit.

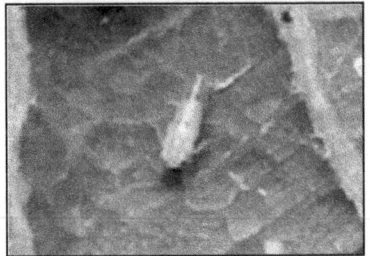

Juice-Grape: A liquid material extracted from grape berries.

Kelvin Temperature Scale: This temperature scale was designed by Lord Kelvin (William Thomson, 1824-1907). The scale contains range from water freezes at 273.16 K and boils 373.16 K. Each unit on this scale is called as Kelvin.

Kniffin: Kniffin is one of the training system used to the framework of the vine, useful to developing canopy in a systematic row. There are three type modern kniffins - 1 four-arm kniffin, six-arm kniffin, Umbrella kniffin.

Lactic Acid: The acid form in wine under anaerobically (extreme oxygen deficiency) condition during the process of malolactic fermentation.

Ladybird Beetle: Small round bright-coloured and spotted beetle that feed on aphids and another insect pest. It is native to Australia, introduced elsewhere to control scales insect. They are usually coloured in some combination of black and red, orange or yellow and often have spots on their wings cover. These insects control the mealy bug infestation.

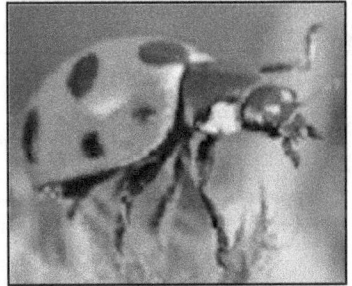

Lamina: The expanded portion or blade of the leaf, attached to petiole above ground organ specialized for photosynthesis.

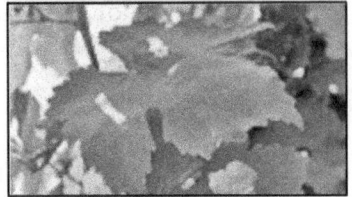

Larva: Newly hatched forms of an insect before they undergo metamorphosis stages (eggs, larva, pupa and adult) contain three distinct body parts, head, thorax, and abdomen. For moth and butterflies, the larval stage is called as a caterpillar.

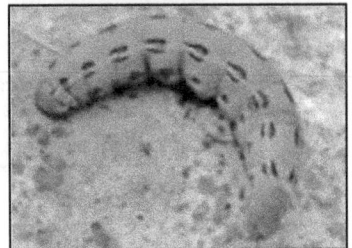

Latent Bud: An axillary bud which remain undeveloped or dormant for a long time, but may eventually grow if a branch break or is cut off just above a latent bud, the bud may develop a new shoot to replace the wood that has been removed.

Lateral: These are the shoots located in the axil of a leaf.

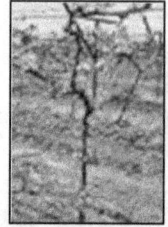

Lateral Shoot: A shoot growing more or less obliquely from the parent shoot and subordinate to it, arising from axillary bud.

http://www.treeterms.co.uk/definitions/lateral-shoot

Laterites and Lateritic Soils: Laterite is a peculiar formation soil in India and some other tropical countries, with an intermittently moist climate. This soil contains mixture of the hydrated oxides of aluminum and iron with manganese oxide, titania.

http://www.yourarticlelibrary.com/soil/8-major-soil-groups-with-statistics-explained/44694/

Leaching: The downward movement of water or soluble nutrients from the soil, due to heavy rain and heavy irrigation. The viticulture term refers to the loss of certain qualities of the soil, such as pH when rainwater removes or "leaches out" carbonates from the soil.

Leaf: The flattened structure of plant part, typically green and blade-like structure that is attached to stem directly via a stalk. This takes part in the vital activity of photosynthesis to produce carbohydrates used for plant growth and development.

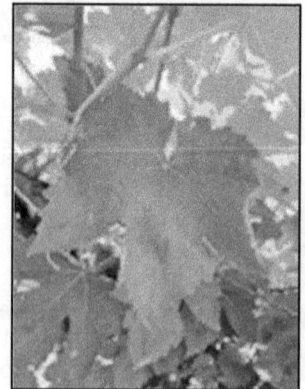

Leaf/Shoot Blight: A plant disease is causing lesions, withering, wilting or death of affected plant parts.

Leaf Blade: It is also known as the lamina, and it is dominant especially in grass leaf or the broad portion of a leaf as distinct from the petiole.

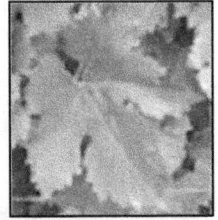

Leaf Blight and Berry Necrosis: In grapes minute yellowish spot on the upper surface of the leaves which later enlarge and turn into brownish patches mostly along the leaf margins. In grapes, leaf blight is caused by *Alternaria alternata*.

Leaf Bud: Leaf bud is an undeveloped shoot, and it is present in the axil of a leaf or at the tip of the stem.

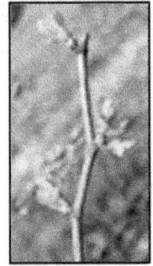

Leaf Hopper: One of the largest families of plant-feeding insects. They feed by sucking the sap of the grape plant and are found almost anywhere cosmopolitan in distribution. They belong to Cicadellidae family. They are a destructive pest for grapes vines. The damage first appears as a scattering of small white spots, with severe infestation and continues feeding the entire leaf turn yellow, finally brown, dries up and drop off.

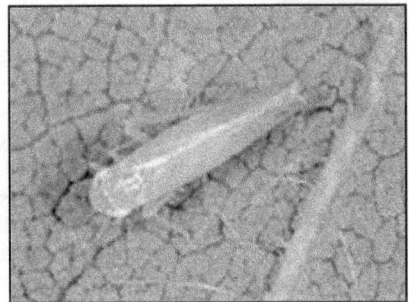

Leaf Scar: The mark left on berries or twig or stem due to chemical or mechanical cause.

Leaf Spot: Round lesion found on the leaves of many species of plant, mostly caused by parasitic fungi.

Lees: It is the dead yeast leaving at the bottom of a fermentation container.

Legs: This is an adherence of wine on the side walls of the glass.

Lenticels: Small, corky pores or narrow lines on the surface of the stem of woody plants that allow the interchange of gases between the interior and the surrounding air mostly.

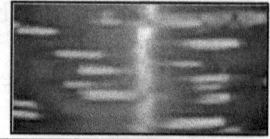

Lesion: These are the lesions of wound marks on the vine.

Life Cycle: The complete succession of changes undergone by an organism during its life.

https://kidskonnect.com/science/life-cycles/

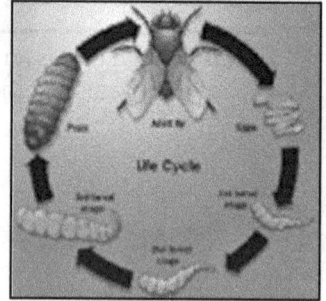

Light Saturation: This is the point at which photosynthesis rate becomes almost nil. This depends on the carbon dioxide and PAR (photosynthetically active radiations).

Light Soil: A term used for sandy and coarse textured soil. Leaching of nutrients is more in light soils.

Light trap: Collection and destruction of adult beetles by handpicking in daytime and setting up of light traps @2-3/ha in the night time during July-August.

The Grape Entomology By Mani M., Shivaraju C., Narendra Kulkarni S.; page 84; 2013.

Lignified: To turn into wood through the formation and deposition of lignin in cell walls.

Lime Sulfur: Lime sulfur is a mixture of calcium polysulfides formed by reacting calcium hydroxide with sulfur, used in pest control. It can be prepared by boiling calcium hydroxide and sulfur together with a small amount of surfactant. It is normally used as an aqueous solution, which is reddish-yellow in colour and has a distinctive offensive odour.

http://www.useranswers.com

Lithosphere: The outer solid part of the earth, including the crust and upper most mantle. The lithosphere is about 100 km thick.

Lotus Illustrated Dictionary of Geology: By Cindy Jones; page;113.Page;113

Lobe: The outer solid part of the earth, including the crust and uppermost mantle. The lithosphere is about 100 km thick Grape leaf lobes are a somewhat rounded section of the blade located on the side of the leaves.

Lysimeter: It is a device used for measuring percolation and leaching losses from a column of soil under controlled condition.

Nitrogen Economy in Tropical Soils; edited by N. Ahmad; page:428;1996

Maceration: The phenolic material of the grapes, tannin, colouring agent (anthocyanin) and flavor compound are leached from the grape skin seeds and stem into the must during wine making process.

http://en.wikipedia.org

Macroclimate: The climate of large area such as region or a country.

Macroscopic: Visible to the naked eyes.

Magnesium: It is secondary nutrient and an essential constituent of chlorophyll. It also helps to the production of sugar. 1-ton grapes contain approximately 0.09 kg/ton Mg.

Magnesium Sulphate: The chemical formula for magnesium sulphate is $MgSO_4.7H_2O$. This compound commonly used to supply magnesium ions to plant through soil and fertilizer mixture or by foliar sprays.

Manuka: A grape dried either in the sun or by artificial means are also called as raisin.

https://www.wordnik.com

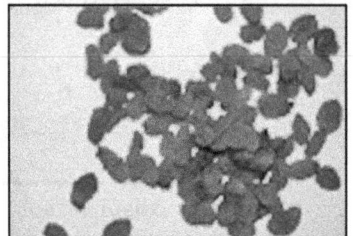

Manures: These are bulky materials such as animal or green manure, which added to improve the physical condition of the soil. This manure decomposed by soil microorganisms.

Margin, Leaf: The boundary area extending along the edges of the leaf.

http://www.cactus-art.biz

Mating Disruption: A technique of insect management designed to confuse male insect for searching mate pheromone into the environment to interfere the reproductive cycle of an insect.

Maturity: The position at which a fruits or vegetative part grows fully developed.

Mealybug: Mealy bugs have become an increasing threat to the grapevine in peninsular India causing a massive loss in the field. Mealy bugs are major pests in Maharashtra, Andhra Pradesh, Tamil Nadu and Karnataka. They feed on the plant sap/juice from growing shoots and berries and act as a vector for several plant diseases.

Mechanical Injury: Crushed, picked, scraped vigorously, or struck due to equipment or other physical activities to any part of the plant is called a mechanical injury.

Metamorphosis: The biological process of transformation from an immature form to an adult form in two or more distinct stages.

http://simple.wikipedia.org

Meteorology: It is a study of atmosphere. Based on weather information we can forecast many things.

Microbial Influence: Microbial biogeography is nonrandomly associated with regional, varietal, and climatic factors across multiscale viticulture zones.

http://www.pnas.org/content/111/1/E139.full

Microclimate: A microclimate is a general atmospheric geographical region where the climate varies from the surrounding area. The sequence of climatic changes within the tiny region.

Microflora: This is a group of bacteria. These bacteria mostly occur on leaf surface.

Micronutrients: Micronutrients are those elements essential for plant growth that are needed in only very small (micro) quantities.

Micropropagation: Growing plant from seeds or small pieces of tissue under a sterile condition in a laboratory on specially selected media under controlled environment and then planting them out.

Mite: A little living creature is belonging to the order acarina related to spiders and ticks. Some mites live freely and others as parasites on plants and animals carry diseases and cause allergies.

http://www.medicinenet.com

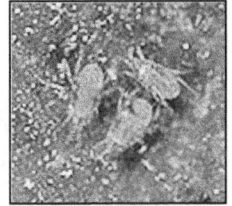

Moisture: This is the water vapor content of the atmosphere. Moisture per cent increases the relative humidity in the air.

Monoculture: Monoculture is agriculture practice of growing same plant year after year over wide area.

http://en.wikipedia.org

Monoecious: A plant having the male and female reproductive structure in separate flower but on the same plant.

http://www.ext.colostate.edu

Monsoon: This is a rainy season in India. It starts from June in India and stays for 3-4 months.

Mottle: A diffuse term for mosaic symptom in plant leaves in which the dark and light green are less sharply defined that is symptomatic of many virus diseases.

http://www.plantpath.cornell.edu/glossary/

Mouthfeel: This is the feel of the aroma and taste of the wine.

Mulch: A layer of material (wood chips, straw, leaves, *etc.*) placed around plants to hold moisture, prevent weed growth and enrich or sterilize the soil.

Environmental Engineering Dictionary, 4th edition; By C. C. Lee; page 511.

Mulches: Mulch is a layer of organic material applied on the surface of the soil to improve soil fertility and health of the soil. Layer also help to prevent unwanted rays comes from sun.

http://en.wikipedia.org

Mummfication: This is shriveling of the berries.This occurs due to overload and imbalance in the nutrition.

Muscadine: This is a grape genotype of *Rotundifolia* species.

Muscat: This is grape variety having Muscat flavour. This variety is mostly is grown in the Tamil Nadu state of India.

Mutation: A sudden change in the genetic structure (DNA) is called as a mutation. This may occur due to the treatment of radiation or mutagenic chemicals.

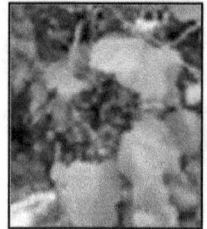

Mycelium: This is a thread like filaments found in fungus.

http://www.thefreedictionary.com/hypha

NADP⁺ (Nicotinamide Adenine Dinucleotide Phosphate): Nicotinamide Adenine Dinucleotide Phosphate is the cofactor for some dehydrogenases enzyme. During photosynthesis in the PS II photoreaction, NADP⁺ is converted into NADH.

Necrosis: This may be a nutritional disorder of localized death of cells or tissues. The death or disintegration of cells or tissue resulting in a darkening of the affected tissue.

Necrotic: These are the dead tissues occurs due to chemical toxicity or disease infection.

Nematicide: A chemical compound or physical agent that kills nematodes.

http://www.plantpath.cornell.edu/glossary/

Nematode: Microscopic worm-like animals that live in soil or water, or as parasites in plants and animals.

http://eviticulture.org/glossary-of-grape-terms/

Nematode Resistant Rootstocks: These rootstocks were designed to provide broad and durable resistance to nematodes to propagate well and have good horticultural characters such as long internodes. More than seven years required for nematode eradication between two successive vines.

Neutral Grape Varieties: White or table grape considered as neutral varieties which category grape has less aromatic character.

Night Soil: Night soil is human excreta, having solid and liquid nitrogenous waste product. It used as compost manure to the enhancement of soil nutrients of the plant. It is a valuable agricultural resource since ancient times.

Nitrogen: Air is the primary source of nitrogen for plant nutrition and the only leguminous plant can directly use this free nitrogen with the help of symbiotic bacteria. Plants derive from the soil their nitrogen in the form of nitrogen and ammonium.

Nitrogen Cycle: Nitrogen is one of the chief and essential constituents of protein and a nucleic acid molecule that play the basic role in cell metabolism, growth, reproduction and transmission of heritable character.

Node: The region of a stem between two internodes, where there is branching of the vascular tissue into leaves or other appendages.

http://www.ucmp.berkeley.edu/glossary/ glossary_8.html

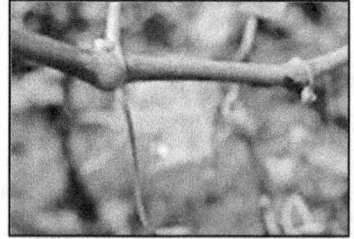

Nymph: Larva of an insect with incomplete metamorphosis.

http://www.uky.edu/Ag/Entomology/ythfacts/4h/ unit2/hoigr and cf.htm

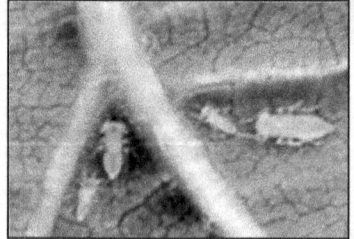

O

Oak: The strongest and most durable wood that lasts for extended periods of time in either wet or dry condition. Commonly used wood source for fermentation vessel and barrel aging.

Obvate: The egg-shape of leaf with the small end towards the stem.

Oenologist: An expert in the field of oenology (all aspects of wine and winemaking) is known as an oenologist.

http://en.wikiversity.org/wiki/

Oenology: The science and study of wine from production to testing.

Organic Agriculture: Organic agriculture is a holistic production management system that promotes and enhances agro-ecosystem health, including biodiversity, biological cycles, and soil biological activity.

http://www.fao.org/organicag/oa-faq/oa-faq1/en/

Organic Matter: This is a decomposed waste material of plant and animal origin.

Organic Viticulture: It is a cultivation of grapes without the use of toxic chemicals.

Organoleptic: This is the wine tasting based on the taste, smell and sight of the wine.

Osmotic Pressure: The phenomenon of a pure solvent to move through a semi-permeable membrane and into a solution containing a solute to which the membrane is impermeable. This process is of vital importance in biology as the cell's membrane is selective toward many of the solutes found in living organisms.

https://www.boundless.com/microbiology/textbooks/boundless-microbiology-textbook/culturing-microorganisms-6/physical-antimicrobial-control-69/osmotic-pressure-406-5527/

Ovary, Plant: In a flowering plant, the part of the flower that encloses the ovules. When the ovary matures, it becomes the fruit.

http://www.ucmp.berkeley.edu/glossary/glossary_8.html

Oven Dried Soil: Soil that has been dried in an oven at 105°C until it reaches a constant weight.

Over cropping: It is an overload of the bunches that results in the little accumulation of the sugar when it is a desirable character of the quality.

Ovicide: A pesticide toxicant for insect or mite eggs, usually restricted to action on eggs of phytophagous insects.

http://www.encyclo.co.uk

Own-Rooted Vine: The vines are grown on their roots and non-grafted.

Oxidation: This is a process of degradation of wine in the presence of air or oxygen. Oxidation of wine plays a vital role in the formation of colour and flavour of final product.

Palate: This implies to flavor, tastes and textures produced by the wine in the mouth.

Paleosols: Soils formed long ago under climate different from those that now exist. Also called as fossil soils.

Panicle: The panicle is a cluster of flowers.

http://www.wisegeek.com

Para heliotropism: Para heliotropism is a movement of leaves to avoid or to minimize absorption of solar rays, to maintain the water loss and leaf temperature.

Paraquat (1,1 Dimethy-4-4' -Bipyridylium) Dichloride): 1,1 Dimethy-4-4' - Bipyridylium) Dichloride) 1,1 Dimethy-4-4' -Bipyridylium) Dichloride is non-selective herbicide used as a post-emergence weedicide to kill all types of weeds. It is a contact herbicide and has little residue toxic to the soil. It kills the weeds by destroying the chlorophyll of plant; care must be taken not to spray on vine leaves.

Paraquat injury: Injury from the contact herbicide paraquat (Gramoxone Extra) typically appears as rusty orange spots or irregular-shaped blotches on leaves.

http://grapes.msu.edu

Parasite: An organism that lives on or within another organism as a host and obtains nutrients from the host without benefiting or killing the host.

Concise Dictionary of Biology, By Editorial Board; page 146.

Parenchyma: A generalized cell or tissue in a plant. These cells may manufacture or store food, and can often divide or differentiate into another kind of cells.

http://www.ucmp.berkeley.edu

Parthenocarpy: Seedless fruit that develop without any fertilization of ovules.

Parthenogenesis: It is a form of asexual reproduction where growth and development of embryos occur without fertilization, *i.e.,* unfertilized egg develops into a new individual.

Partial Root-zone Drying: It is a technical system for vineyard irrigation where only a section of a vine root system received measured amounts of water. It is potentially useful to reduce vine water use, reduce canopy vigor, and maintain crop yields and fruit quality as compared with standard irrigation methods.

Participatory Technology Development: Technological development building on existing farmer's knowledge and enhancing farmer's capacities to experiment and to communicate their finding.

Particle Density: The density of the soil particles, the dry mass of the particles being divided by the solid volume of the particles, in contracting with bulk density. Its unit is $g\,cm^{-3}$.

Particle Size Analysis by Sieving: The division of a sample by sieving into size fractions. These are simple and rapid chemical tests of soils designed to give an approximation of the nutrients available to plants. Interpretation of results depends upon previous standardization of soil tests with the response to fertilizers. It varies for different kinds of soil.

Pasteurization: Is a process of heating food, which is usually a liquid to a particular temperature for a predefined length of time and then immediately cooling it after it is removed from the heat.

http://furkan.uz/en/equipments-for-dairy-plants-uzbekistan-tashkent/equipment-for-dairy-plants-in-uzbekistan-tashkent-milk-processing-equipment-for-dairy-plant/milk-products/milk-pasteurization/

Patents: A patent is a governmental grant; issued in the India or other country to an Inventor, or his/her assignee, the right to exclude all others from making, using, offering for sale, importing or selling the Invention within the government's territory or to other country to inventor for particular time period.

Pathogen: Any microorganism that causes diseases is called as a pathogen. (bacteria, fungus, viruses, yeast, molds, and parasites).

http://www.infoplease.com/ipa/A0762206.html

Pathogenesis: The term describe the origin and development of the disease, and whether it is acute, chronic, or recurrent.

http://en.wikipedia.org

Pathogenicity: Pathogenicity is the ability of a pathogen to cause a disease.

http://textbookofbacteriology.net

Pectic Enzyme: An enzyme that added to fruit juice prior to the fermentation process to enhance the clarification process. Pectic enzyme destroys haze causing pectin cells that can leave the wine with a permanent milky appearance known as a "pectin haze".

Pectins: The gel like group of polymer substances found in the cell wall of the grape plant. Pectin is produced by fruit in the ripening process and helps the ripe fruit berries to remain firm.

Pedicel: A pedicel is a stem that attaches single flower to the main stem of the inflorescence.

http://en.wikipedia.org

Pedology: Pedology is a branch of soil science which deals with the survey, genesis, classification of soil development and soil formation for land use planning.

Peduncle: The stem or branch that holds a group of pedicels is peduncle.

http://gardener.wikia.com/wiki/Pedicel

Penicillium Rot or Blue Mould Rot: Penicillium species incite this rot and rarely becomes a post-harvest problem because they can infect through wounds only. High relative humidity favors the disease.

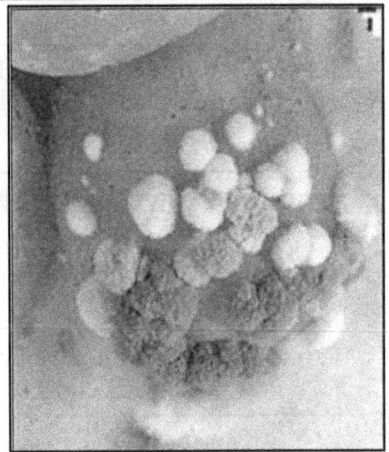

Percolation, Soil Water: It is a downward movement of water through the soil. Especially the downward flow of water in saturated or nearly saturated soil at hydraulic gradients of the order of 1.0 or less.

Concept Dictionary of Agricultural Sciences; By I. C. Gupta, S. K. Gupta; Page no. 356

Perennial Weed: A weed reproduces through seed, but also through various vegetative structures. Mainly vegetative growth in a first year and set seed in the second year.

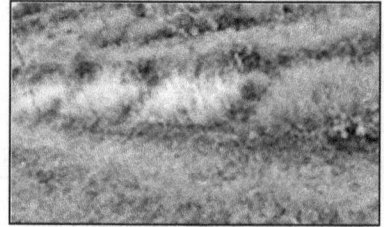

Perennial Wood: This is a woody growth of the vine which is retained year after year.

http://eviticulture.org/glossary-of-grape-terms/

Perfect Flower: A flower that contains male and female parts.

http://eviticulture.org/glossary-of-grape-terms/

Pericarp: This is a ripened ovary consist of endocarp, mesocarp and exocarp.

Periderm: The outer core layer of a plant that replaces the epidermis of primary tissue. Cells have their walls impregnated with cutin and suberin.

http://wordinfo.info

Perlite: Granular volcanic rock used to improve the aeration in potting soil but no nutrient value.

http://www.encyclo.co.uk

Permanent Wilting Point: This is the condition of soil at which plant cannot recover its turgidity.

Encyclopedia of Agriculture and Food Systems: Neal K. Van Alfen; 5-volume set, Volume 1, page 47

Permeability, Soil: Permeability is the measure of the soil's ability to penetrate water flow through its pores.

Pest: An insect, rodent, nematode, fungus, weed or another form of terrestrial or aquatic plant or animal life that is injurious to health or the environment.

Environmental Engineering Dictionary; 4 th edition; By C. C. Lee; Page 580.

Pesticide: Any substance (chemical or microbial) which by virtue of its toxicity is used to control harmful organism. It includes bactericide, fungicide, herbicide, insecticide, nematicide or rodenticide, *etc.*

http://www.encyclo.co.uk

Petiole: The slender stem that supports the blade of a foliage leaf.9

http://www.thefreedictionary.com/petiole

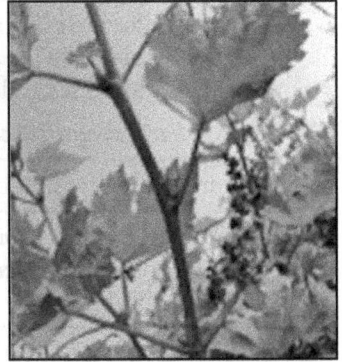

pH: It is a measure of how acidic/basic an aqueous solution is. This is a negative logarithm of H ion concentration.

Phenology: The study of the periodic (seasonal) phenomena of plants and animal life cycle and their relation to the weather and climate (*e.g.*, the time of flowering plants).

http://www.encyclo.co.uk

Phenotype: External appearance or observable characteristics of the plant or living organism.

Pheromone: These are the chemical substances secreted by insects used to attract the opposite sex.

Phloem: Living conducting tissue of a plant's system composed of sieve tube companion cells, fibers and sclereids to convey the products of photosynthesis, particularly sucrose, from the leaves to growing tissues.

http://pecan.ipmpipe.org/glossary

Phosphrous: Phosphorus is a macronutrient. It encourages the formation of the new cell, promotes root growth and hastens leaf development. Essential for transport of energy through the plant, especially green parts (leaves, clusters).

Photoperiodism: Photoperiodism is the physiological reaction of plants to the length of day or night. It is a biological response to a change in the proportions of light and dark on a 24-hour daily cycle.

http://www.cliffsnotes.com/sciences/biology/plant-biology/growth-of-plants/photoperiodism

Photosynthates: Photosynthates are the synthesized chemical substance used as food energy to fuel plant growth and maintain plant function.

Photosynthesis: It is the process of formation of sugar by the plant from carbon dioxide and water in the presence of light.

Photosynthetic Pigments: A substance that absorbs light, often selectively carotenoids, xanthophyll, chlorophyll was some examples of photosynthesis pigment.

Phylloxera: Phylloxera small, sap-eating greenish insect of the genus Phylloxera, closely related to the aphid. Phylloxera feed on leaves and roots of grape vines, causing significant damage.

http://arlindo-correia.com/060904.html

Phylum: A category in the hierarchy of classification by class and kingdom.

Physiological Disorder: Disorders caused by the disturbances in the normal metabolic function of the vine created by an unknown or complicated factor that are dealt under this head.

Phytophagous: Plant eating pests.

Phytotoxic: Harmful to plants.

Pinching: It is a form of pruning that is done forcontrolling, shaping and directing the growth of foliage.

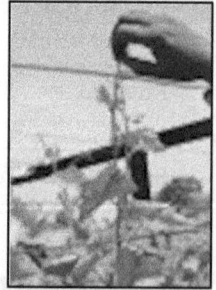

Pink Berries: This is a formation of pink color on the skin of the berries at version stage in non coloured grape varieties.

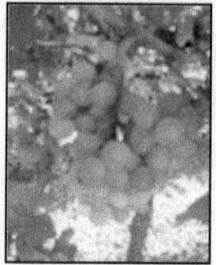

Pistil: It is female reproductive organ of a flower.

Pistillate Flower: It is a female flower having only pistils and lacks stamens.

Pith, stem/root: The ground tissue occupying the center of the stem or root within the vascular cylinder; usually consists of parenchyma.

http://www.fda.gov/Drugs/DevelopmentApprovalProcess/FormsSubmission Requirements/ElectronicSubmissions/DataStandardsManualmonographs/ucm071788. htm

Plant Analysis: It is analytical procedure use to determine the nutrient contents of the plants or plant parts.

Plant Bark: It is the outermost tough protective covering of the stems of trees, shrubs and vines.

http://eviticulture.org/glossary-of-grape-terms/

Plant Breeders' Rights: Plant Breeders' Rights is applied only to plant. It is a patent-like system that allows the plant variety owner to prevent anyone doing unauthorized uses of the variety.

Plant Food Ratio: The ratio of the number of fertilizer units in a given mass of fertilizer expressed in the order NPK. This ratio may be based on nitrogen as unit.

https://www.iso.org/obp/ui/#iso:std:iso:8157:ed-1:v1:en

Plant Growth Microorganism: Microorganisms make available different kinds of nutrient plant, which are required for their growth. (i) they must be able to colonize the root (ii) they must promote plant growth.

Plant Hair/Trichromes: Plant hair is prolonged epidermal cell; on a grapes stem, leaves, root. Hairs on plants are extremely variable in their presence across species, the location of plant organs, density (even within a species), and therefore functionality. Plant hairs may be unicellular or multicellular.

Plant Injury: Plant injury would be considered mechanical damage to the external plant part. Plant damage caused by an insect, pest, animal, chemical or environmental agents.

Plant Pathology: It is the study of plant diseases and the abnormal conditions that constitute plant disorders.

http://www.plantpath.cornell.edu/glossary/

Plant Physiology: Plant Physiology is the study of life activities, responses and functions of plants.

Plant Respiration: Respiration the process in which the stored energy within food is released for the plants use is called respiration. Respiration involves the reaction of oxygen with carbon and hydrogen.

Plant-Incorporated Protectants: These are substances that act like pesticides produced and used by a plant to protect it from pests such as insects, viruses, and fungi.

The Promise of Biotechnology; Page 40, oct 2005.

Pollen: Fine particles containing the fertilizing element of plants (male) formed by anther of plants.

http://lethamshank.co.uk/glossary/glossary.php?letter=P

Pollen Crossing: Crossing is a simple technique of cross-pollination in that; a matter of brushing the pollen onto stigmas of the emasculated cluster by manually.

Pollen Germination: This is a germination of seeds and spores in plants, fungi, and bacteria respectively.

Pollination: Pollination is a transfer of pollen from the male reproductive organ (stamen) to the female reproductive organ (pistil) of the same (self-pollination) or of another flower (cross pollination).

http://www.infoplease.com/encyclopedia/science/pollination.html

Polyculture: Various species with different nutritional and environmental requirements are cultivating together; competition should be avoided.

Polyphageous: These are the organisms that feeds on various food sources.

Polyploidy: This term refers to a numerical change in chromosomes.

http://en.wikipedia.org

Pomace: The solid residue (skin, seeds and stems) left behind by draining juice from white must, or draining new wine from a red fermentation tank.

http://www.napavalley.edu

Post Harvesting Technology: Post harvest technology is inter-disciplinary "Science and Technique" applied to the agricultural product after harvest for its protection, conservation, processing, packaging, distribution, marketing, and utilization.

Potassium: Potassium is essential element by grapevine especially for fruit clusters development. Potassium also balances water flow in the plant. It also helps in maturation of canes.

Potassium Sorbate: Potassium sorbate is the potassium salt of sorbic acid. Its primary use is a food preservative; it is effective in a variety of applications including wine stabilizer and preservative it inhibits mold and yeast growth.

Potential Acidity: Potential acidity defined as the acidity developed due to adsorbed hydrogen and aluminum on the soil colloids. The magnitude of this potential acidity is very high.

Powdery Mildew: A devastating fungal disease of grape vines that, unlike most fungal diseases, thrives in dry climates. Also called odium, it occurs in most of the grape growing regions of the world. This is the most troublesome fungus disease of grapes. It can be controlled by timely application of fungicide and sulfur dust directly onto the vine leaves and immature fruit.

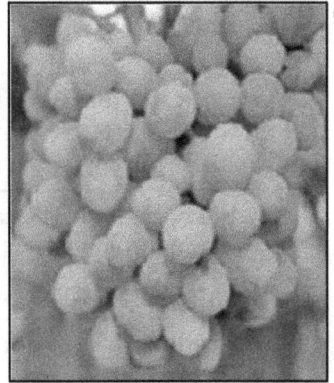

PPV and FR Act: It is an attempt by the Indian Government to recognize and protect the rights of both commercial plant breeders and farmers in respect of their contribution made in conserving, improving and making available plant genetic resources for development of new crop varieties and to encourage the development of new plants varieties. For agricultural development and protection of plant breeder right act has been established in India; in 2001.

http://www.lakshmisri.com/Uploads/MediaTypes/Documents/plant_variety_protection_in_india.pdf

Prebloom: The stage before flowering in grapes.

Precision Viticulture: This is the cultivation of grapes with the use of modern tools and techniques to obtain maximum quality yield.

Pre-Emergent Herbicide: Herbicides that are applied to the soil before the crop emerges and prevent germination or early growth of weed seeds *i.e.*, that kills plants as they germinate.

http://en.wikipedia.org

Pressing: Extracting juice or wine from the skin, pulp and seeds using wine press.

Primary Bud: The primary bud is the main fruiting bud and contains 2-3 inflorescence primordia and 6-12 leaf primordia by the time of winter dormancy, depending on the variety and species.

http://wine.wsu.edu

Primary Inoculum: The primary inoculum is the part of the pathogen (that is bacteria or fungal spores or fungal mycelium) that over winter (over-seasons) and causes the first infection of the season, known as primary infection.

http://ipm.ncsu.edu/apple/chptr4.html

Primordia: The first recognizable, histological differentiated stage in the development of an organ.

http://www.thefreedictionary.com/primordium

Primordium: These are the cells from which organs form and further it leads to growth.

Prior Art: A prior art can include patents, publications, documents, written articles, devices known, on sale, or used by the public, *etc.* A prior art is any evidence that your invention is already known.

Productivity, Soil: The capacity of the soil to produce a particularly characterized plant or sequence of plants under a specified environmental condition.

http://nesoil.com/gloss.htm

Propagation: To increase number of plants by sexual or asexual reproduction.

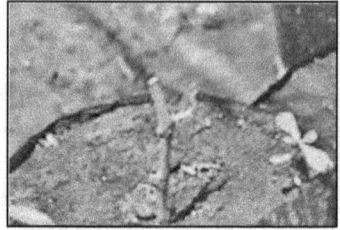

Protectant: This is a chemical that protect the infection of the pathogen on the plant parts.

Protonymph: This is the first developmental stage in many acarids.

Pruning: This is the cutting of shoots for the reemergence of growth for specific purposes.

Pubescent: Covered with fine, short, soft hairs. In plants may affect wetting of foliage and retention of spray.

http://www.encyclo.co.uk

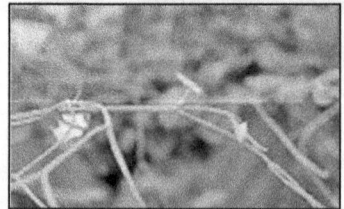

Pulp: The soft inner mass and juicy part of the grape interior or grape berry.

Pupa: It is a stage of cocoon formation and inside this cocoon larva undergoes a transformation and forms an adult.

Pyrazines: A group of aromatic nitrogen containing cyclic compounds; they are the source of intense "green" aromas in the grapes, and they are indigenous to Bordeaux.

http://www.slideshare.net/8802566877/winemaking-glossary-ofterms

Q

Quarantine: Restriction on the movement or existence of something (plant, animals, any material or people's activities) brought under regulation. This is done to prevent or limit the introduction or spread of a pest or control or eradicate a pest already introduced. This practice reduces or avoids losses that would otherwise occur through damage done by the pest or through the continued cost of control measures.

Quick Soil Tests: These are simple and rapid chemical test of soils designed to give an approximation of the nutrient available to plant.

Dictionary of Soil Science; By Subhash Chand; page 73.

R

Rachis: The skeleton of branched stems that gives a grape bunch or cluster its shape.

http://www.lasirenawine.com/Joie-de-Vivre/Glossary

Racking: Decanting clear juice or wine from above the sediment (lees) in a container. This is the easiest method for getting rid of solids that have settled to the bottom in a tank or barrel. Wine containers commonly have a built in "racking valve" placed 20 inches (~half a meter) above the base valve for use in racking wines during production.

http://www.edenwines.co.uk/Glossary_r.html

Raisin: Raisin is the dried grapes also called as manuka.

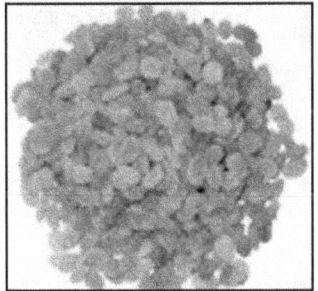

Receptacle: The enlarged tip of the stem to which the sepals, petals, stamens, and pistils are usually attached. In some accessory fruits, for example, the pome and strawberry, the receptacle gives rise to the edible part of the fruit.

Recombinant DNA Technology: Genes from one species are introduced into a non-related organism. This technology used in biotechnology, medicine, and research.

Reference Laboratory: Laboratory of recognized scientific and diagnostic expertise for a particular animal disease and testing methodology; includes the capability for characterizing and assigning values to reference reagents and samples. 125

http://www.oie.int/doc/ged/D7710.pdf

Refractometer: A hand-held device used to measure the sugar content of grapes; the measurement is based on the proportional diffraction of light by sugar in grape juice. It helps to determine ideal harvesting times of grapes so that the product arrives in an ideal state to consumers or for subsequent processing steps such as vinification.

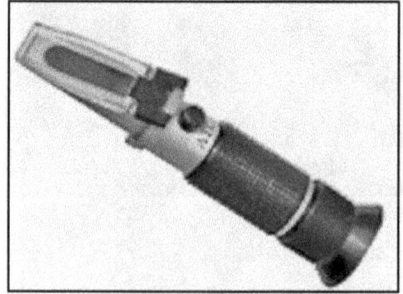

Relative Humidity: The ratio, expressed as a percentage, of the amount of water vapor present in a given volume of air at a given temperature to the amount required to saturate the air at that temperature. It is usually expressed in percentage.

Electrical Power Transmission and Distribution: Aging and Life Extension technique, By Bella H. Chudnovsky; Page No. 158.

Remote Sensing: Remote sensing is the acquisition of information about an object or phenomenon without making physical contact with the object and thus in contrast to in situ observation. The technique employs such devices as the camera, lasers, radio-frequency receivers, radar system, sonar, seismographs, gravimeter, magnetometer, and scintillation counters.

Renewal Spur: In grapevines the renewal spur allow for replacement of growth or renewal of growth. The purpose of renewal spur is to give options in selection of fruiting canes in future years.

Reniform Nematode: Nematodes are microscopic roundworms found in many habitats. The nematodes most damage the secondary and the feeder roots. Most active in warm weather in moist. As a result, the nutrient uptake is affected, and the vine appears sick.

Residual: This is the left over material of the chemicals used for efficient control of insects or pathogens.

Residual Effect of Manure: This refers to the residual beneficial effect of application of farmyard manure on the succeeding crops. This beneficial effect is due to improvement in the physical condition of the soil, and also due to unutilized plant nutrient.

Residual Sugar: The term commonly used in wine analysis referring to the content of unfermented sugar in wine already bottled. In a dry wine, this primarily involves the non-fermentable sugars arabinose and rhamnose.

Resistance: The ability of an organism to exclude or overcome, completely or in some degree, the effect of a pathogen or other damaging factor.

http://www.plantpath.cornell.edu/glossary/

Retardation Factor: The capability of a soil for slowing or retarding the movement of a solute and is defined for solutes subject to equilibrium reactions with the soil matrix.

Soil and Environmental Science Dictionary; edited by E.G., Gregorich, L. W. Turchenek, M.R. Carter - 2001; page no. 301.

Rhizopus Rot or Nested Rot: Caused by Rhizopus stolonifer (Ehrenb. Fr.) or R. arrhizus Teshe sp., this fungus causes rot usually. Rotted areas on the berries are soft and brown, drip juice under humid conditions and may be covered with small spherical dark sporangia emerging through cracks in the skin of diseased berries or borders of a wound.

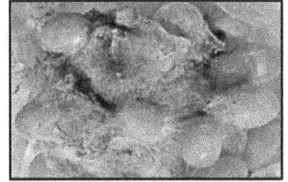

Ripening: The process of maturation of berries into a particular level of the sugar, acid and pH with the purpose of harvesting. Series of physiological changes occur in the berries that indicating ripening of grape started.

Root - lesion Nematode: In heavier soils infected young vine remain feeble, often fail to establish a root system and eventually die. Below-ground symptoms on roots distinctly show lesions, which are initially brown and later turn black. In severe infection, black lesions combine and griddle the roots. Infected plants have reduced root system resulting in a reduction in potassium and zinc uptake.

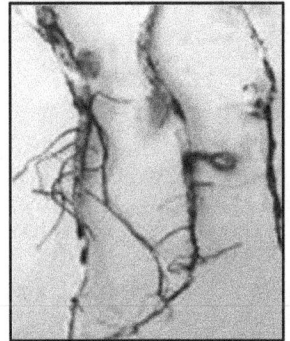

Root Rot: It is fungal disease; occurring on the single vine that are excessively irrigated. The affected plant loses their luster, turn pale and finally dry.

Root Tip: The root tip is the lower portion of tissues on the tip of the root; contain the root cap and meristem tissue.

Root-Knot Nematode: Root-knot nematode larvae infect plant roots exhibit severe galling. Galling is the result of the proliferation of cells of the affected roots. The vines show stunting and poor growth. Young shoots remain short and chlorotic. In a severe attack, the vines get defoliated.

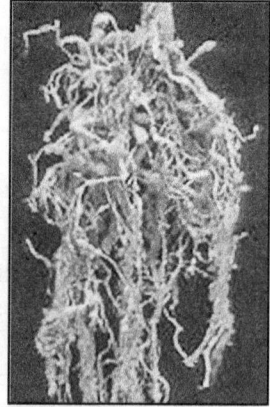

Rust: This is a fungal disease caused by Phakopsora vitis (syn *Physopella vitis, P. ampelopsidis*). Rust affected leaves show spots correspond to a yellow-orange mass of spores. Among the commercial cultivars, it is an important disease of Bangalore Blue in south interior Karnataka.

http://www.dpi.nsw.gov.au/__data/assets/pdf_file/0009/ 458334/Exotic-Pest-Alert-Grapevine-leaf-rust.pdf

Saccharometer: A saccharometer is a device that is used to measure the concentration of sugar in a solution. It is Wine making a tool that uses specific gravity to measure the sugar content of the grape juice.

Salination: The process in which soluble salt accumulate in the soil.

Salt Creek: This is a rootstock of grapes that is known to restrict the uptake of chloride.

Sap: A watery solution of sugars, salts, and minerals that circulates through the vascular system of a plant.

http://www.thefreedictionary.com/sap

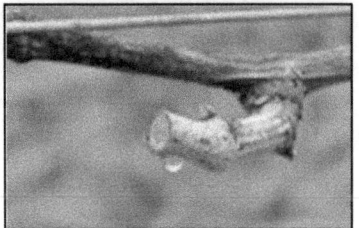

SARD (Sustainable Agriculture And Rural Development): This means that agriculture and rural development become sustainable; when they are ecologically sound, economically viable, socially just, culturally appropriate and human based on a holistic scientific approach.

http://www.inmotionmagazine.com/global/sard.html

Scab: A hyperplastic symptom characterized by rough, crusty lesions formed by excessive cork production. A disease in which such lesions form.

http://succulent-plant.com/glossary

Scale Insect: These are the tiny and immobile insects which suck the sap from leaves, petioles, and shoots and also bunches. As a result of an attack by scale insects, the vines become weak. A Severe scale infestation results in the death of branches and decline of plant yield.

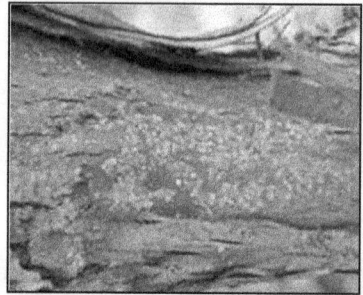

Sclerotium: Hard, resistant, multicellular resting body, usually with a differentiated cortex and medulla that under favourable conditions can germinate to produce mycelium or sexual or asexual fruiting bodies. (*Pl. sclerotia.*).

http://www.plantpath.cornell.edu/glossary/

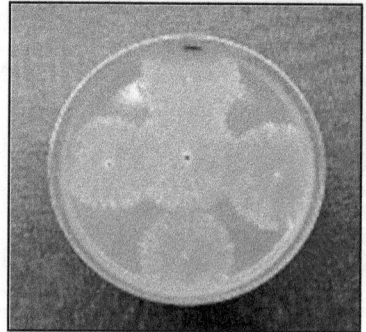

Scorch: Burning of leaf margins due to lack of sufficient water, excessive transpiration or injury to the water-conducting system of the plant.

Second Crop: This is a harvesting of yield one after another.

Secondary Bud: In viticulture - the second largest of the three buds in a compound winter bud, generally may or may not be fruitful and does not develop unless there is damage to the primary bud (*e.g.*, frost injury).

Secondary Fermentation: In Winemaking, any fermentation that happens after the primary (yeast) fermentation has been completed. Malo-lactic is a secondary fermentation that occurs in most red, and some white, still wines. Another secondary is the yeast fermentation that is used to change still wine into sparkling wine.

http://eviticulture.org/glossary-of-grape-terms/

Secondary Inoculum: Propagules produced as a result of infection by the primary inoculums is secondary inoculum.

Secondary Metabolite: Secondary metabolites are organic compounds that are not directly involved in the normal growth, development, or reproduction of an organism.They are considered as end products of primary metabolism, *viz.* Alkaloids, phenolics, essential oils, steroids, *etc.* In plants, secondary metabolites are believed to be important in the attraction of pollination and to have a function in defense, as some metabolites have antimicrobial properties.

Secondory Mineral: A mineral resulting from the decomposition of a primary mineral or the re precipitation of the product of decomposition of a primary metal.

http://eusoils.jrc.ec.europa.eu/ESDB_Archive/glossary/Soil_Terms.html

Seedy fruit: Fruit having many seeds.

Seedless fruit: When fruits are entirely devoid of seeds, contain a much-reduced number of seeds or present aborted seeds.

http://www.ncbi.nlm.nih.gov

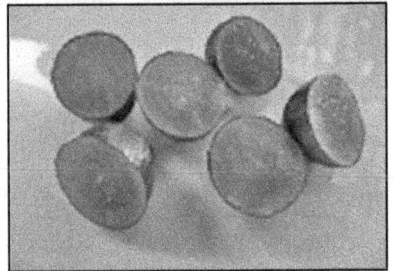

Selective Herbicide: These are the agrochemicals also known as weed killers. Selective herbicides kill specific targets while leaving the desired crop relatively unharmed. Herbicides used to clear waste ground, industrial sites, railways and railway embankments are not selective and kill all plant material with which they come into contact. Smaller quantities are used in forestry, pasture systems, and management of areas set aside as wildlife habitat.

http://www.aristobiotech.com/products

Self-Pollination: This is a kind of pollination that can occur when a flower has both stamen and a carpel (pistil) are present and in which the cultivar or species is self-fertile, and the stamens and the sticky stigma of the carpel contact each other in order to accomplish pollination.

http://en.wikipedia.org

Senescence: Senescence refers to all of the changes that take place in a plant that will finally lead to the death of cells, tissues, and, eventually, the whole plant body. These changes can be seen to occur in some cells even in very young, vigorously growing plants. For example, the contents of those cells that make up the xylem tissue must senescent and die very early in development. Senescence can also be seen in large, multicellular plant organs including leaves and fruits. Golden fields of ripening grain and the reds and yellows of the fall landscape in forests are both due to the pigment changes occurring during the early stages of senescence in millions and millions of leaves. This senescence are often accompanied by changes in the levels of plant hormones in the cells, with shifts in the absolute amount and sensitivity towards the gaseous hormone, ethylene, playing a pivotal role.

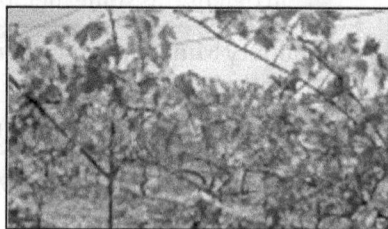

Sepal: Sepal, a modified leaf, part of the outermost of the four groups of flower parts. The sepals of a flower are collectively called the calyx and act as a protective covering of the inner flower parts in the bud. Sepals are usually green.

http://encyclopedia2.thefreedictionary.com

Serrations: Serrations are the tooth-like structures at the margins of the leaf.

Shading, berry: Artificial shading of grapes decreased berry size, flavonoids, skin tannin, and nor-isoprenoids.

Shatter: The drying up of a high percentage of unsuccessfully pollinated pistils leaving a nearly bare stem skeleton (with few berries attached).

http://www.lasirenawine.com/Joie-de-Vivre/Glossary

Shoot Thinning: It is a removal of excess shoots to facilitate better fruit quality and production.

Shoot Topping: This refers to the removal of the apical portion of shoot to increase berry size and quality of the produce.

Shoot Vigour: This is the vigorous growth of the shoot, and it can be determined by shoot thickness, internodal distance, leaf size and shoot growth.

Shot Berry: A few small, seedless grapes found in an otherwise normal bunch of wine grapes. The cause is improper fertilization, during the bloom period.

http://www.eldoradohomewinemaking.com/index.php/wine-glossary

Shot Hole borer: It is a minor pest in certain pockets of grape growing areas in peninsular India and feeds mainly on dry wood. Initially, the beetles bore and make pin holes into the main trunk to a depth of 2-3 cm, and later they bored randomly in all the directions.

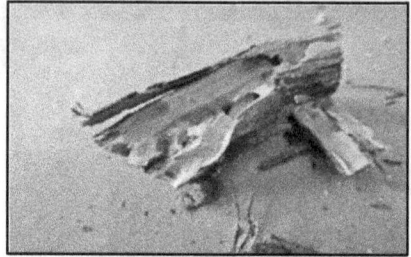

Shouldered Cluster: This is a cluster present on the branched rachis.

Signs: Signs are the external look and physical evidence of the damages caused by biotic or abiotic factor of the plants abnormality.

Simazine(2-Chloro- 4,6 Bis (Ethylamino) S-Triazine): It is 2-Chloro- 4,6 Bis (Ethylamino) S-Triazine used for the weed control chemical in the vineyard. This chemical has no adverse effect on vine growth, yield or quality. The impact of this chemical persists for a long period after application.

Sink: The movement of carbohydrates occurs from where they are made by light reaction (photosynthesis) to where they will be stored or used, called a sink." For example, during ripening, a leaf is the source of whole photosynthate fruits are a sink.

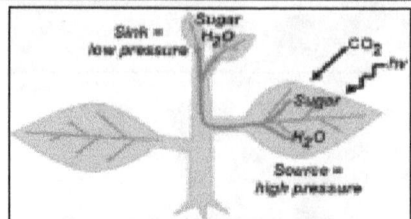

Sinus: This is the angle between the lobes of the leaf blade.

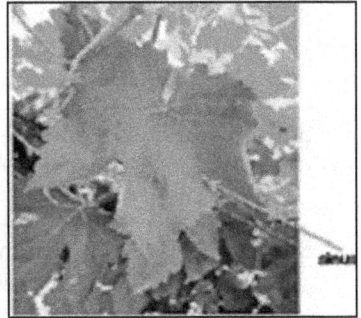

Skin Thickness: The thickness of the skin means; it contains natural waxy coating on the upper surface of the fruit. Thin grape skin is a desirable character in raisin grapes from the view of faster drying and eating quality.

Soft-Wood Cuttings: These are the green shoots used for vegetative propagations.

Soil Additive: Properly mixing of qualified external additive with soil layer to improve texture and condition of a soil is called as soil additive.

Soil Auger: A tool for boring into the soil and withdrawing a small sample for field or laboratory observation.

Soil and Environmental Science Dictionary; edited by E.G. Gregorich, L. W. Turchenek, M.R. Carter - 2001; page no. 331.

Soil Buffering Capacity: Soil's ability to maintain a constant pH level during action on it. The effectiveness of soil buffering systems depends on numerous physical, chemical, and biological properties of soils.

Soil Classification: The systematic grouping of soil types according to their properties (physical, chemical, biological) classes into one or more categories or level.

Soil Colour: Soil colour depends upon soil composition; the development and distribution of colour in soil results from chemical and biological weathering, especially redox reactions minerals, protein, oxide, organic matter. Colour can be used as a clue to the mineral content of a soil. Prior to selecting and planting grapes, conduct a

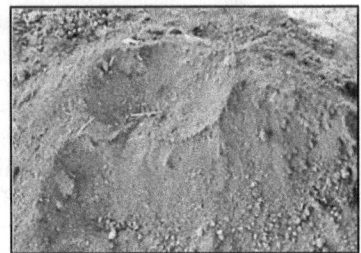

thorough soil test; analysis of soil pH, physical characteristics, organic matter content, and chemistry.

Soil Conservation: Soil conservation is an effort made by man to prevent soil erosion from acidification, Stalinization or other chemical soil pollution to retain the fertility of the soil.

Soil Enzymes: The total biochemical activity of soil is comprised of a series of the reaction catalyzed by enzymes. More than 52 proteins have been shown to be active in the soil, including oxidoreductase, hydrolase, and transferase.

Soil Erosion: Soil erosion is the rapid blowing or washing of soil from its natural place in a landscape to another location, usually many feet or possibly even miles away.

Soil Extract: The separation of mineral, vitamin, element from a solution of soil by using a different process like filtration, centrifugation, suction, or pressure.

Soil Fertility: The ability of the soil to produce crops having economic value and protect or keep the capacity of the soil for future crop production.

Soil Formation Factor: The variable, usually interrelated natural agencies that are active in and responsible for the formation of soil. The factors are usually grouped into five broad categories like- parent rock, climate, organisms, topography, and time.

Soil Fumigant: A Chemical substance incorporated or spread into the soil to kill insects, nematodes, and other soil born diseases that damage growing plants, seeds, *etc.*

Soil Genesis: The branch of soil science which deals with the study of soil formation.

Soil Genesis, Classification Survey and Evaluation, Volume 1; By A. K. Kolay; page no 2.

Soil Loss Tolerance: The maximum soil erosion loss that is offset by the theoretical maximum rate of soil development which will maintain equilibrium between soil losses and gain. 136

https://www.soils.org/publications/soils-glossary/536

Soil Management: It is a branch of soil science that deals with the study of tillage and planting operation, cropping practices, fertilizer, lime, irrigation, herbicides and insecticides application.

Soil Mineral: Soil minerals play a vital role in soil fertility. A natural inorganic compound with definite physical, chemical, and crystalline properties that occurred in the ground.

Soil Morphology: The physical constitution, particularly the structural features, of a soil profile as exhibited by the kinds, thickness and arrangement of the horizons in the profile, and by the texture, structure, consistency, and porosity of each horizon.

http://www.ecologydictionary.org/Soil_Morphology

Soil of Indian Arid Ecosystem: The arid zone of India located in a northwestern part of the country mostly covers Western Rajasthan, Southern Punjab and Haryana and Northern Gujarat.

Soil Organic Matter: The organic fraction of the soil that includes plant-animal residues at various stages of decomposition, cells and tissues of soil organism and substances synthesized by soil organisms.

Nurturing the Soil-feeding the People; By Winfried. Scheewe; page 267.

Soil Productivity: Productivity is the present capacity of a soil to produce crop yield under a defined set of management practices. It is measured in terms of the yield about the input of production factors.

Dictionary of Soils and Fertilizers; By L. L. Somani, vol-4, part-5.

Soil Science: Science that deals with soil as a natural resource on the surface of the earth including soil formation, classification and mapping, physical, chemical, biological, and fertility properties of soils per se; and these properties in relation to the use and management of soils.

http://zandcti.in/wp/s/Soil_science.html

Soil Structure: The arrangement and organization of primary and secondary particles in a soil mass are known as soil structure. Soil structure controls the amount of water and air present in the soil. Soil particles may be present either as single individual grains or as an aggregate *i.e.* group of particles bound together into granules or compound particles. These granules or compound particles are known as secondary particles. 140

http://www.agriinfo.in

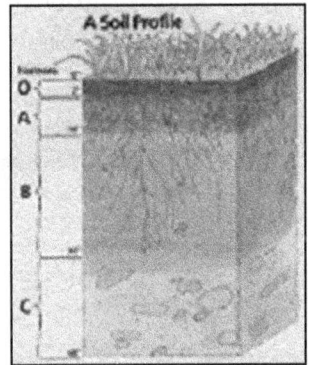

A Soil Profile

Soil Test: A chemical, physical, or biological procedure that estimate a property of the soil.

Soil Texture: Soil is comprised of different sizes of mineral particles. The relative amounts of each of these various sized particles constitute a soils texture. There are three main soil textures; sandy, silty or clayey and there are pluses and minuses to each one.

http://gardening-time.blogspot.in

Soil Variant: A kind of soil that differs enough from recognized series to justify a new series name but is so limited in area that creation of a new series is not justified at a given time.

Dictionary of Soils and Fertilisers By L. L. Somani, vol4, part-5

Soil Water: Soil water is part of the global hydrological cycle, soil stores water is of great importance to crop production and the vitality of the land. Soil water content on a volume basis (qv = volumetric water) is the ratio of water volume (Vw) to total soil volume (Vt).

Source: The cells where photosynthesis occurs and glucose is made known as the source (leaf cells).

Specific Gravity: The specific gravity scale on a hydrometer is used to measure the sugar content and sweetness of juice or wine.

Sporangiophore: A specialized fungal branch bearing one or more sporangia.

https://www.wordnik.com

Sporangium (Pl.Sporangia): A unicellular or multicellular sac-like structure in fungi that produces asexual spores through meiosis.

Spore: A microscopic reproductive cell of non-flowering plants (*i.e.* ferns, lichens, mosses, fungi, and algae), capable of developing into adult.

Spur: A shortened stub of cane, usually formed by pruning the cane to a length of only two to four nodes (buds).

http://www.marylandwine.com

Stabilizer: A chemical such as potassium sorbet that is added to wine before they are sweetened to maintain it in a stable or unchanging state by disrupts the reproductive cycle of yeast. Yeasts present are unable to reproduce, and their population slowly diminishes through attrition.

Stamen: Stamen is the male reproductive organ of a flower. It consists of the anther and the slender filament that holds it in position.

http://www2.ca.uky.edu

Staminate Flower: A flower possessing only stamens (male parts) but no Pistils (female parts).

Stem Borer: Stem borer is a slender, elongated beetle with a light brown or yellowish body. Stem borer has been reported from Maharashtra, Andhra Pradesh, Karnataka and Tamil Nadu causing damage to grapes in the field.

Stem girdle: It is major pest reported in Maharashtra, Punjab, Tamil Nadu and Andhra Pradesh. Adult beetles have powerful mandibles used to girdle around the young branches, main stem, at any place from 15 cm to 3 meters above the ground level at night.

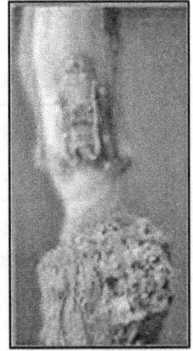

Stigma: The part of the female flower that which receives the pollen from the anthers.

Stock: To avoid the harmful effect of the soil, certain wild species are used and on which the desired scion is grafted. These are called as rootstock in grapes.

Stomatal Regulation: Stomatal movement depends on guard cells pressure. Stomatal cell opening and closing depends upon potassium and chlorine concentration. Guard cells chloroplasts fix CO_2 photosynthetically to form sugar. Which contribute to the solute buildup required for stomatal opening. Environmental factors that affect stomatal movement.

Stomata: Tiny openings on the undersides of grape leave through which pass gasses and water. The principal gas that passes through stomata is carbon dioxide, which is on its way in to get captured by the chlorophyll and be turned into sugar. 120

http://www.slideshare.net/8802566877/winemaking-glossary-ofterms

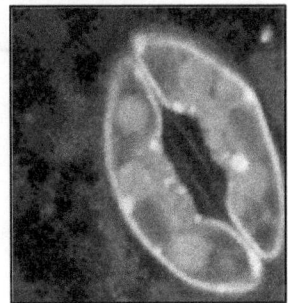

Straight Fertilizer: Qualification gave to a nitrogenous, phosphatic or potassic fertilizer having a declarable content of one primary nutrient only. 145

http://nptel.ac.in/courses/103107086/module1/lecture1/lecture1.pdf

Strategic Leaf Pruning: For aeration within the canopy, leaves are removed which is called as strategic leaf pruning.

Stratification: Storing of seeds at low temperatures under moist conditions in order to break dormancy. Stratification requires cold (usually around 40°F), moist (but not wet) conditions for a period of time (usually 3-8 weeks).

Style: This is a stalk through which pollen tubes grow to reach the ovary.

http://en.wikipedia.org

Stylets: The slender, hollow, piercing and sucking organs of insects and nematodes that feed on plant sap.

https://insects.tamu.edu/students/undergrad/ento402/Glossary/Glossary_S.html

Suberin: A mixture of fatty acid derivatives deposited in cell walls to make them impermeable to water, as in corky cells. Often produced in wound healing reaction and to prevent penetration by a pathogen.

http://www.encyclo.co.uk

Succulent: A plant that can store water in its tissues and then withdraw it during times of drought. Water storage tissue may be found in the stem, leaves, or roots depending on the species.

http://palaeos.com/plants/glossary/glossarySi.html

Sucker: These are the side shoots growing at the trunk base on the rootstock in grafted vine of grapes.

Suckering: This is a process of removing suckers.

Sugar-Acid Ratio: This is a grape quality parameter and indicates the palatability of the berries. It is also called as brix acid ratio.

Sulphur Bacteria: Sulphur bacteria convert sulphide into sulphuric acid. These compounds act on insoluble soil compound like calcium carbonate, magnesium carbonate, calcium silicate, tricalcium phosphate and bring them into the soluble state.

Sunscald: Due to exposure to bright sunlight and excessive heat plant parts get damaged mainly berries is called sunscald.

Surfactant: A material that spreads along a surface, changing the properties of the surface by reducing the surface tension.

Susceptibility: This is an inability of an organism to resist the attack of pathogen or pest or environmental conditions.

Susceptible: Inability to resist attack of pathogen.

Swabbing: The process in which a multi-class mixture of pesticides or other material is covered on the surface of grape leaves or whole plant by using swab.

Systemic: These are the chemicals or pathogens that translocate through the cells in the plant system.

Systemic Fungicide: These are the fungicide that enters/translocate in the plant cells.

Systemic Herbicide: A compound that is translocated readily within the plant, either from the foliar application down to roots or from soil application up to leaves and has an effect throughout the entire plant system.

'T' Trellis: This is a type of training system. This is also called as the telephone system.

Tannin: A group of simple and complex phenol, polyphenol, and flavonoid compounds, bound with starches, and often so amorphous that they are classified as tannins simply because at some point in degradation they are astringent and sometimes bitter in taste. Tannin in wine comes from grape skins, stems, or seeds (if seeds are crushed or broken open by mistake) and from wood contact during barrel aging. Seed tannin is the least desirable in wine because this type of tannin is usually quite bitter. Tannin is the component that allows red wine to age, acting as a natural preservative, helping the development and balance of the wine. Tannin is potent antioxidants and promotes color stability.

Tartaric Acid: The tartaric acid most prominent natural acid of grapes. This acid not usually found in other fruits or vegetables.

https://supervalu.ie

Tatura System/Tatura Trellis: This is a training system with the diagonal orientation of the canopy. This looks like an X shape.

Teinturier: An ancient red-wine grape whose flesh and juice is red due to anthocyanin pigments accumulating within the pulp of the grape berry itself, rather than just the skin.

Teinturier Varieties: These are the grape varieties having coloured pulp and skin.

Telephone System: This is a grapevine training system having 'T' shaped look and looked like telephone wires and pole.

http://www.nhb.gov.in/Horticulture per cent 20Crops/Grape/Grape1.htm

Temperate Climate: Climates with distinct winter and summer seasons, typical of regions found between the Tropics of Cancer and Capricorn and the Arctic and Antarctic Circles. Considered the climate of the middle latitudes.

http://www3.weathertrackcast.us

Temperature Inversion: This is the situation in which air temperature near the ground is coolest, and it increases with the altitude.

Tendril: A threadlike often branched appendage on a stem or leaf that coils around plants or other objects to provide support for a climbing plant.

http://www.npwrc.usgs.gov/resource/plants/vascplnt/glossary.htm

Tensiometer: Instrument for measuring the moisture content of the soil. A tensiometer works by measuring the amount of tension that the soil exerts on the instrument as it loses soil moisture and water is removed from the instrument.

http://www.extension.org

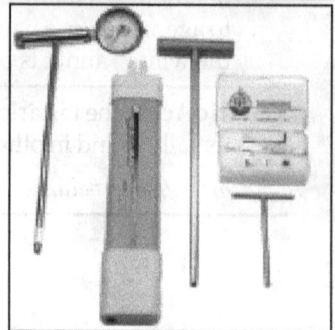

Terpene: This indicates the aroma of the wine, and it is a class of unsaturated hydrocarbons.

Tertiary Bud: This is a third bud and develops after primary and secondary buds are killed.

Tetraploid: It is four sets of chromosomes in plant.

http://agridr.in/tnauEAgri/eagri50/GBPR111/lec20.pdf

The Grape Flower: The grape petals are fused into a green structure termed the calyptra but commonly referred as the cap. The grape flower does not have conspicuous petal. A flower consists of a single pistil and five stamens.

http://www.extension.org/pages/31097/parts-of-the-grape-vine:-flowers-and-fruit#.VP6LE8liJkg

Thinning: This is a removal of excess shoots or berries to achieve quality grape production.

Threshold: This is a point at which action is required to prevent the loss.

Thrips: These are the tiny insects having the sucking type of mouthpart. They are considered as a major pest of grapes.

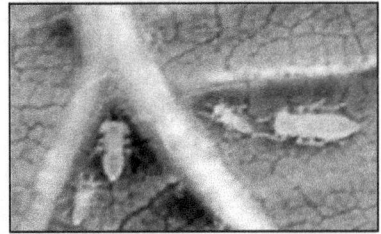

Tillage: The mechanical manipulation of soil activity or process of preparing land for growing plant.

Encyclopedia of Soil Science; R. Lal, page 1773

Tissue Analysis: It is an analysis of any tissue sample for various purposes like nutrient analysis and its composition.

Tissue Culture: A very sterile practice of propagating plants from the mother plant. Done in laboratory conditions. Orchids, hosta, and daylilies are done by this method.

http://www.emilycompost.com/garden_glossary.htm

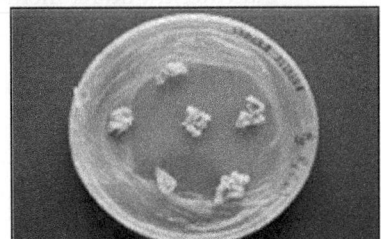

Titratable Acidity: Titratable acidity is only a measure of hydrogen ion consumed by titration with a standard base to a defined end point, usually pH 8.2. Titratable acidity depends on both the amount of acid present and the pH. Expressive of the level of acid present in the juice. In Europe, expressed as g/l as sulfuric acid. In the U.S., in g/l or, more often, g/100 ml, as tartaric acid. Value range between 0.3 and 1.3 g/100 ml.

http://www.napavalley.edu/

Tobacco Caterpillar: This pest is of common occurrence in Maharashtra and Hyderabad. Active during August-September. Light brown with dark spotted wings; Larvae colour varies depending on leaves and makes the leaf surface papery.

Tomentum: It is a short, soft pubescence or a covering of fine, soft hairs.

http://en.wikipedia.org

Top Dressing: An application of fertilizer to a soil surface, without incorporation, after the crop stand, has been established.

Top working: It is a changing of cultivar by grafting onto old or unfruitful vines.

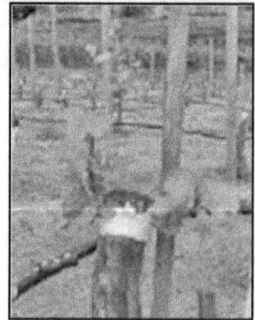

Torus: This is a hook like structure inside the berries also called as flower stalk. This indicates the attachment of the berries.

Trade And Service Marks: A word, name, symbol, or device (or any combination because of that) that an organization adopts to identify its goods or services and distinguish them from the goods and services of others.

http://www.basicpatents.com/tmlawsor.htm

Traditional Agriculture: Traditional agriculture is the ancient and most commonly practiced form of farming all over the world. Soil is the natural supporting and nourishing medium for the cultivation and growth of plants.

Training, Grapevine: This is an orientation of vine shoots in such a way that it can utilize the maximum sunlight for the highest production of grapes.

Trait: A genetically distinguishing quality characteristic or condition.

http://www.yourdictionary.com/trait

Translocation: Movement of water and nutrients from one part of a grapevine to another through the xylem and phloem.

http://eviticulture.org/glossary-of-grape-terms/

Transpiration: Loss of water from a vine by evaporation through tiny pores (stomata) in the leaves.

Trellis: The physical support structure used for training vines. The structure is usually made of wood or metal for vigorous vines.

http://eviticulture.org/glossary-of-grape-terms/

Trunk: The central, vertical structure of a grapevine which supports all the top growth.

http://www.edenwines.co.uk/Glossary_t.html

Turgor Pressure: Turgor pressure is the force that is exerted on a plant's cell wall by the water (in the cytoplasm) within the cell.

U

Ultra Violet Spectrophotometry: An analytical instrument used to measure the concentration of compounds in grapes, or other samples, based on the absorbance of light. Spectrophotometry can be used to measure sugars, acids, YAN, phenolics, and other compounds in grapes.

http://www.vitisgen.org/glossary.html

Uneven Ripening: This is the phenomenon in which some berries are ripened, and some are unripe at harvest stage.

V

Vacuole: A vacuole is a fluid-filled, membrane-bound cavity inside a cell. It may be a reservoir for fluids that the cell will secrete to the outside, or may be filled with excretory products or essential nutrients that the cell needs to store.

http://www.probertencyclopaedia.com/browse/BV.HTM

Value-Added Products: Product can be considered value-added if the original agriculture raw product modified, changed or enhanced to increase the value of the newly developed product.

Varietal Character: The characteristics specific to a variety. Varietal characteristics include flavor, aromatic compounds, acidity levels, thickness of skin and size of the individual grapes.

Variety: A taxonomic rank below subspecies in botany, varieties are usually the result of selective breeding and diverge from the parent species or subspecies in relatively minor ways.

Encyclopaedic Dictionary of Biology; By S. Choudhary; page 513.

Vascular Elements: These are the conducting cells in plants *i.e.*, xylem and phloem.

Vector: It is an organism that transmits disease from one organism to another.

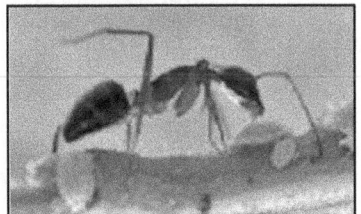

Vegetative Propagation: Propagation without pollination by way of separating vegetative parts (*i.e.*, branches, stolons, buds) from the mother plant and planting them, so they take root and grow.

http://www.mondofacto.com/facts/dictionary? vegetative+propagation

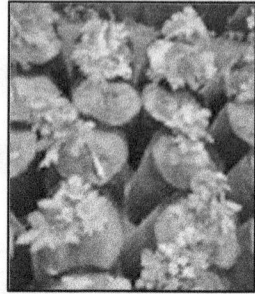

Vein: These are the vascular tissue of the leaf and are located in the spongy layer of the mesophyll cells.

http://en.wikipedia.org

Ventral: In plants, pertaining to the surface facing the stem to which the leaf or other structure is attached. In animals, it refers to the body surface facing the ground.

Glossary of Plant Biology and Biochemistry, http://www.shieldsgardens.com/info/ Glossary.html

Veraison: The onset of ripening. The first colour change from green to purple on the vine (in black grapes) or green to yellow green (in white grapes), accompanied by a softening of the texture. This is the first step in the maturation of grapes on the vine.

http://en.wikipedia.org/wiki/Veraison

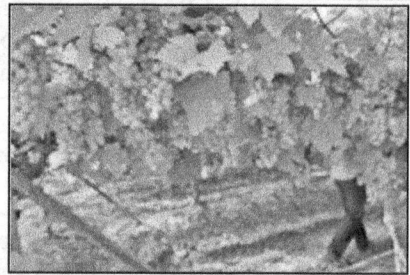

Vergin Soil: Never cultivated soil.

Vermitechnology: Vermitechnology (VT) is the application of earthworm in the development of arable soils, break down of plant organic matter, aeration and drainage. It also helps to monitoring soil fertility, organic and heavy metal, non-biodegradable pollutant. In this way, they are helpful in managing waste biomass.

Vertical Shoot Positioning (VSP): This is the orientation of the shoots in upward direction called as VSP.

Vigor: A measure of the quality of growth expressed by a grape vine, as opposed to capacity that measures the quantity of growth and development.

http://www.edenwines.co.uk/Glossary_v.html

Vine Density: This is the number of grape plants per unit area.

Vine Shelter: These are the plastic tube used to protect the young plants.

Vine Size: This is a sum of growth of the vine in a particular season.

Vine: A vine is a plant that needs support as it grows. Some vines grow by twining around other objects for support.

http://www.enchantedlearning.com/subjects/plants/glossary/indexv.shtml

Viniculture: This is science and art of making wine also called as enology.

http://www.allfoodbusiness.com/wine_savvy.php

Vinifera: Scientific name of the primary species of Vitis (vines) used for winemaking *Vitis vinifera* produces nearly all the world's wines.

http://www.edenwines.co.uk/ Glossary_v.html

Vintner: A wine merchant.

http://www.merriam-webster.com/dictionary/vintner

Virulence: It is a measurement of pathogenicity of the pathogen.

Virulent: This is a state of severe disease caused by pathogen.

Virus: It is a microscopic parasite, often cause disease. It can be occur in plants, animals to bacteria.

Virus-Tested, vines: These are vines subjected to virus testing by following laboratory procedures.

Viticulture: The science, art and study of grape growing.

http://www.thefreedictionary.com/ viticulture

Vitis labrusca: This is an American grape variety commonly called as 'Concord' used for juice, jams, jellies and sweet wine.

Vitis rotundifolia: This is American variety also called as muscadines.

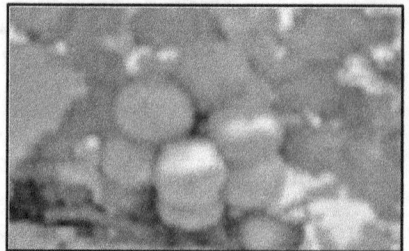

***Vitis vinifera*:** This group consists of about 60, mostly hardy, deciduous, woody vines, commonly known as the Grape. These traditional natives of North America, China and Japan, are mainly valued for their delicious fruits, which can be eaten fresh or dried, squeezed as juice or fermented as wine.

http://www.botany.com/vitis.html

Voracious: Consuming, or eager to consume, high amounts of food.

http://www.thefreedictionary.com/voracious

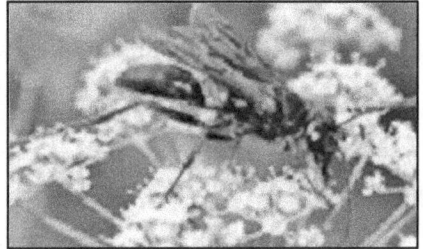

Wasps: Wasps feed on half ripe and ripe berries causing extensive damage. An insect of the order Hymenoptera, usually with a hard shiny body, usually with biting mouthparts, and a predatory or parasitic lifestyle. Parasitic wasps are increasingly used in agricultural pest control as they prey mostly on pest insects and have little impact on crops.

Water berry: This is a physiological disorder where the berries become watery and having less sugar at the ripening stage.

Water Cycle: Water is constantly being cycled through the atmosphere, ocean and land to the sky and back again on earth surface. This process, known as the water cycle. Water never leaves the earth.

http://creditunionloans.info/the/the-water-cycle-science-education-at-jefferson-lab.html

Water Harvesting: It means storing rainwater where it falls or capturing the runoff for a productive purpose like drinking, irrigation, increase groundwater recharge, reduce storm water discharges and urban swamping. Water harvesting potential = Rainfall (mm) x collection efficiency.

Water Retention Curve: A graph showing the soil water content versus applied tension or suction.

http://jan.ucc.nau.edu/~doetqp-p/courses/env320/glossary.htm

Water Sprout: These are basal shoots on the rootstock.

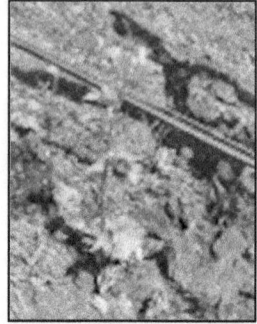

Waxy Or Yeast Rot: It is caused by *Geotrichum candidum*. Affected Berries showed soft tan coloured area and covered with dirty white wax colonies of the fungus. The rotting is fast and fermented juice evident typical odour.

Weed: These These are the plants out of place and not intentionally sown and compete with the main crop for nutrient and water.

Weed Control: Weeds are controlled by a number of preventive cultural technique like suitable crop rotation, green manures, pre-drilling, mulching and by mechanical, or chemical methods.

Wettable Powder: It is a dry formulation of agrochemical when mixed with water it forms a clear solution that can be used for spraying.

Whip and Tongue Grafting: This is a method of grafting in which both stock and scion are cut diagonally, and their surfaces are provided with matching tongues that interlocked when the graft is tied.

Oxford Dictionary of English edited by Angus Stevenson, page no 2022.

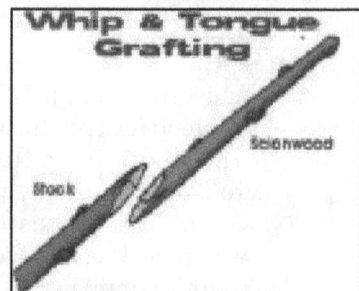

Whip & Tongue Grafting

Scionwood

Shook

Wildlife Control: Use of netting, noise producing mechanical or electrical devices or other methods by grape growers to protect the crop from wildlife damage.

Wilt: A disease (or symptom) characterized by a loss of turgidity in a plant (*e.g.*, vascular wilt). 84

http://www.plantpath.cornell.edu/glossary

Wine grape: A grape used in making wine especially a European grape (*Vitis vinifera*).

Wing: This is well developed lateral branch of rachis from the bunch.

http://eviticulture.org/glossary-of-grape-terms/

Winkler Scale: This is the scale developed by Winkler to evaluate the region's heat summations.

Wire Moving: On training system, it is an act of positioning of foliage wire to orient the shoots.

Woody part of vine: The exterior of woody parts of the vine is protected by periderm, which comprises cork cell and is covered by an outer bark consisting of dead tissues.

http://gencowinemakers.com/docs/Grapevine per cent 20Structure per cent 20and per cent 20Function.pdf

Xenotransplantation: Transplantation of cells, tissues, or organs from one species to another.

http://en.wikipedia.org

Xeric: The Very dry environment condition is called as Xeric. It is calculated by Comparing with medium water conditions (mesic) and very wet conditions (hydric).

Xylem: The part of the vascular system that moves water and minerals through the plant.

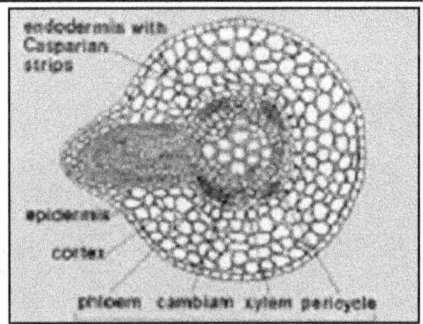

Yeast: A primarily unicellular fungus that divides by budding (or fission) and possesses a glucan or mannoprotein cell wall (budding yeast may contain some chitin around the bud scar). Yeast can occur naturally in the air, especially in areas where fruits are grown. Commonly known as baker's yeast.

Yeast Assimilable Nitrogen (YAN): YAN is a combination of free nitrogen, ammonia, or ammonium ion. Compounds utilized by yeast for the synthesis of proteins, cell wall components, and enzymes during wine fermentation.

Yield: It is a quantity of grapes produced per unit area at harvest.

Zinc: The zinc present in Zn^{2+} ionic form is readily absorbed by leaves. It is essential for healthy leaf development, shoot elongation, pollen development, and fruit set.

Zinc Deficiency: Zinc is essential for uniform fruit maturity and seed formation. Deficiencies are occasionally encountered in vineyards, especially on sandy soils; where vines have been grafted onto nematode-resistant rootstocks. Deficiency symptoms include early appearing leaf chlorosis, reduces leaf size wider petiolar sinus, poor fruit set and berry development and straggly clusters.

Zymology: It deals with the biochemical processes involved in fermentation, with yeast selection, physiology and with the practical issues of brewing.

http://www.memidex.com

References

http://eviticulture.org/glossary

http://www.mcgeescrossroadsweather.com/wxterms.php

Ency. Dic. Of Bioinformatics and Biotech.(Set 2 Vol),By D.J. Atary, page no.2; 2004

http://en.wikipedia.org/wiki/Glossary_of_plant_morphology

http://www.careerindia.com/courses/unique-courses/what-is-agribusiness-scope-career-opportunities-http://en.wikipedia.org/wiki/Glossary_of_winemaking_terms

http://www.ncirossallpointfleetwood.co.uk/weather/terms.htm

http://www.egeis-toolbox.org/documents

http://www.thefreedictionary.com/argol

Pure Appl.Chem.,Vol.78, No.11, pp.2075–2154, 2006

http://www.tifton.uga.edu/lewis/glossary.HTM

http://bofduniverse2.blogspot.in/2010_12_01_archive.html

http://nrcca.cals.cornell.edu/soil/CA2/CA0212.1-3.php

http://bugs.bio.usyd.edu.au/learning/resources/PlantPathology/glossary.html

http://en.wikipedia.org/wiki/Backcrossing

http://eviticulture.org/articles/page/112/

http://www.ceago.com/biodynamic/

http://www.blueplanetbiomes.org/world_biomes.htm

http://www1.lsbu.ac.uk/water/enztech/biosensors.html

Principles of Biochemistry and Biophysics; By Dr. B.S. Chauhan page 780;2008.

http://ghr.nlm.nih.gov/glossary=biotechnology

http://www.seslisozluk.net/danca/Brix

http://grapestomper.com/wineglossary.html

http://pdf.usaid.gov/pdf_docs/PA00K8NJ.pdf

http://xoap.weather.com/glossary/c.html

http://www.italki.com/question/125746

Catena (Impact Factor: 2.48). 06/1998; 32:155-172;Richard Huggett

http://eusoils.jrc.ec.europa.eu/ESDB_Archive/glossary/Soil_Terms.html

Glossary: Irrigation, Drainage, Hydrology And Watershed Management:By R. T. Thokal, A.G. Powar, D.M. Mahale, R.T. Thokal, D. M. Mahale, A. G. Powar; page No.49.;2004

http://www.extension.org/pages/31951/chaptalizing#.V

http://www.nkhome.com/kestrel/weather-resources/weather-glossary/

Integrated Organic Farming Handbook;By Dr. H. Panda; page.16.

http://sis.nlm.nih.gov/enviro/iupacglossary/glossaryd.html

http://en.wikipedia.org/wiki/Dormancy

http://en.wikipedia.org/wiki/Glossary_of_viticulture_terms

http://dictionary.reference.com/browse/drizzles

http://www.cals.ncsu.edu/course/ent425/tutorial/economics.html

http://agsci.psu.edu/elearning/course-samples/TURF_434/Ln_1/L1_4.htm

http://naldc.nal.usda.gov/download/42085/PDF

http://forest.ap.nic.in/GlosTech-E.htm

Crop Post-Harvest: Science and Technology, Crop Post-Harvest: Principles and.edited by Peter Golob,

Graham Farrell, John E. Orchar; page 514.

http:///jid/journal/v133/n9/full/jid2013287a.html

http://www.biologyreference.com/Co-Dn/Differentiation-in-Plants.html

Definitional Glossary Of Agricultural Terms:By Dinesh Kumar; page

http://www.indiaweather.in/Glossaries/gloss_e.aspx

http://beta.krishiworld.com/html/farm_management1.html

Soil and Environmental Science Dictionary;E.G. Gregorich, ýL. W. Turchenek, ýM.R. Carter – 2001

Dictionary of Soils and Fertilizers, By L. L. Somani; Vol 4, part 2 page-505.

http://www.netlibrary.net/articles/Foliar_feed

http://www.mdpi.com/1424-8220/12/8/10759/htm

http://www.ucmp.berkeley.edu/glossary/gloss5/biome/aquatic.html

http://dirtywordsgarden.com/vocab/

http://www.mdtgrow.com/pruningterms.html

http://www.hindawi.com/journals/jb/2012/135479/

http://entnemdept.ifas.ufl.edu/walker/ufbir/chapters/chapter_06.shtml

http://umaine.edu/blueberries/factsheets/weeds/237-glyphosate-for-weed-control-in-wild- http://www.hort.cornell.edu/reisch/grapegenetics/breeding/crossing1.html

http://www.the-gift-of-wine.com

http://www.thevintnervault.com

http://symposium.surry.edu/schedule/category/Viticulture/past/

http://www.eckraus.com/blog/difference-crushing-pressing-grapes

http://dictionary.babylon.com/greenhouse_effect/

Soils: Genesis and Geomorphology-By Randall J. Schaetzl, Sharon Anderson, page No- 760.

http://www.bis.org.in/sf/fad/FAD72538.pdf

http://www.nhb.gov.in/Horticulture per cent 20Crops/Grape/Grape1.htm

http://www.memidex.com

Definitional Glossary Of Agricultural Terms:By Dinesh Kumar; Y. S. Shivay; vol 1 page-130

http://www.fws.gov/habitatconservation/nwi/wetlands_mapping_training/module2/DW4.html

https://www.rankwise.net/www.igpb.in

http://translate.academic.ru/environment per cent 20indicator/en/ru/1

http://www.treeterms.co.uk/definitions/lateral-shoot

http://www.yourarticlelibrary.com/soil/8-major-soil-groups-with-statistics-explained/44694/

https://kidskonnect.com/science/life-cycles/

The Grape Entomology;By Mani M., Shivaraju C., Narendra Kulkarni S.; page 84; 2013.

http://www.useranswers.com

Lotus,Illustrated Dictionary of Geology;By Cindy Jones; page;113.

Nitrogen Economy in Tropical Soils; edited by N. Ahmad; page:428;1996

http://en.wikipedia.org

https://www.wordnik.com

http://www.cactus-art.biz

http://www.supplewine.com

http://www.medicinenet.com

http://www.ext.colostate.edu

http://www.plantpath.cornell.edu/glossary/

Environmental Engineering Dictionary, 4th edition;By C. C. Lee; page 511.

http://www.claibornechurchill.com/files/Harvest_Terms.pdf

http://www.winecoolersandmore.com/content-pages/page-wine_glossary

http://www.linkapedia-wine.com/topics/wine/mutage/40554581

http://www.uky.edu/Ag/Entomology/ythfacts/4h/unit2/hoigr and cf.htm

http://en.wikiversity.org/wiki/

http://www.fao.org/organicag/oa-faq/oa-faq1/en/

https://www.boundless.com/microbiology/textbooks/boundless-microbiology-
    textbook/culturing-microorganisms-6/physical-antimicrobial-control-69/
    osmotic-pressure-406-5527/

http://www.encyclo.co.uk

http://www.wisegeek.com

http://grapes.msu.edu

Concise Dictionary of Biology; By Editorial Board; page 146.

http://furkan.uz/en/equipments-for-dairy-plants-uzbekistan-tashkent/equipment-
    for-dairy-plants-in-uzbekistan-tashkent-milk-processing-equipment-for-dairy-
    plant/milk-products/milk http://www.infoplease.com/ipa/A0762206.html

http://textbookofbacteriology.net

http://gardener.wikia.com/wiki/Pedicel

Concept S Dictionary Of Agricultural Sciences; By I. C. Gupta, S. K. Gupta; Page no
    356

http://wordinfo.info

Encyclopedia of Agriculture and Food Systems: Neal K. Van Alfen; 5-volume set,
    Volume 1,page 47

Environmental Engineering Dictionary; 4 th edition; By C. C. Lee; Page 580

http://pecan.ipmpipe.org/glossary

http://www.cliffsnotes.com/sciences/biology/plant-biology/growth-of-plants/
    photoperiodism

http://arlindo-correia.com/060904.html

http://www.fda.gov/Drugs/DevelopmentApprovalProcess/FormsSubmission
    Requirements/ElectronicSubmissions/DataStandardsManualmonographs/
    ucm071788.htm

https://www.iso.org/obp/ui/#iso:std:iso:8157:ed-1:v1:en

The Promise of Biotechnology; Page 40, oct 2005.

http://lethamshank.co.uk/glossary/glossary.php?letter=P

http://www.napavalley.edu/people/gvierra/Documents/Fundamentals_of_Enology_Class/http://www.lakshmisri.com/Uploads/MediaTypes/Documents/plant_variety_protection_in_india

http://www.darapri.it/glos2in.htm

http://wine.wsu.edu

https://brewdudes.wordpress.com/beer-making-terminology/

http://ipm.ncsu.edu/apple/chptr4.html

http://nesoil.com/gloss.htm

http://m.supervalu.ie

http://www.slideshare.net/8802566877/winemaking-glossary-ofterms

http://www.wordsense.eu/quaffing_wine/

Dictionary of Soil Science; By Subhash Chand; page 73.

http://www.lasirenawine.com/Joie-de-Vivre/Glossary

http://www.edenwines.co.uk

http://www.oie.int/doc/ged/D7710.pdf

Electrical Power Transmission and Distribution: Aging and Life Extension technique, By Bella H. Chudnovsky; Page No 158.

http://www.dpi.nsw.gov.au/__data/assets/pdf_file/0009/458334/Exotic-Pest-Alert-Grapevine-leaf-rust.pdf

http://www.inmotionmagazine.com/global/sard.html

http://succulent-plant.com/glossary

http://www.ncbi.nlm.nih.gov

http://www.aristobiotech.com/products

http://encyclopedia2.thefreedictionary.com

http://www.eldoradohomewinemaking.com/index.php/wine-glossary

Dictionary of Soils and Fertilisers By L. L. Somani, vol4, part-5.

Soil Genesis, Classification Survey and Evaluation, Volume 1; By A. K. Kolay; page no 2.

https://www.soils.org/publications/soils-glossary/536

http://www.ecologydictionary.org/SOIL_MORPHOLOGY

Nurturing the Soil-feeding the People; By Winfried. Scheewe; page 267.

http://zandcti.in/wp/s/Soil_science.html

http://www.agriinfo.in

http://gardening-time.blogspot.in

http://www.hintsandthings.co.uk

http://www.marylandwine.com

http://www2.ca.uky.edu

http://nptel.ac.in/courses/103107086/module1/lecture1/lecture1.pdf

https://insects.tamu.edu/students/undergrad/ento402/Glossary/Glossary_S.html

http://palaeos.com/plants/glossary/glossarySi.html

http://www3.weathertrackcast.us

http://www.npwrc.usgs.gov/resource/plants/vascplnt/glossary.htm

http://agridr.in/tnauEAgri/eagri50/GBPR111/lec20.pdf

http://www.emilycompost.com/garden_glossary.htm

http://wine.gg/glossary/summary/page_12/

http://www.basicpatents.com/tmlawsor.htm

http://www.basicpatents.com/tmlawsor.htm

http://www.yourdictionary.com/trait

http://www.vitisgen.org/glossary.html

http://www.probertencyclopaedia.com/browse/BV.HTM

http://www.surf4wine.co.uk/glossary.html

Encyclopaedic Dictionary of Biology; By S. Choudhary; page 513

http://www.mondofacto.com/facts/dictionary?vegetative+propagation

http://www.enchantedlearning.com/subjects/plants/glossary/indexv.shtml

http://www.allfoodbusiness.com/wine_savvy.php

http://www.merriam-webster.com/dictionary/vintner

http://creditunionloans.info/the/the-water-cycle-science-education-at-jefferson-lab.html

http://jan.ucc.nau.edu/~doetqp-p/courses/env320/glossary.htm

Oxford Dictionary of English edited by Angus Stevenson, page no 2022.

http://pubs.acs.org/doi/abs/10.1021/bk-1998-0714.ch013

https://www.erowid.org/chemicals/alcohol/alcohol_article2_ winemakers_manual.pdf

http://gencowinemakers.com/docs/Grapevine per cent 20Structure per cent 20and per cent 20Function.pdf

Myrmecol News 11:191-199, Vienna, August 2008

http://en.wikipedia.org/wiki/Andisols

http://srpgrapes.com/pest-and-disease.html

http://minecraft.wikia.com/wiki/Biomes

Other References

A Handbook of organic farming; Arun K. Sharma.

Handbook of agriculture ICAR.

The Grape - Improvement, production, and post-harvest management; K.L. Chadha; S.D. Shikhamany.

Grapevine structure and function - by Edward W. Hellman.

http://www.biotechnologyonline.gov.au/topitems/glossary.html

http://www.epa.gov/oecaagct/ag101/cropglossary.html

http://www2.kenyon.edu/projects/farmschool/addins/glossary.htm

https://www.integratedbreeding.net/courses/plant-breeding-concepts-and-methods/index-id=003.php.html

https://www.rpi.edu/dept/finance/docs/research/GlossaryIntellectual Property.pdf

http://www.aromadictionary.com/articles/wineglossary_m-z_article.html

Other Publications of NRC for Grapes.